Applied Mathematical Modelling of Engineering Problems

Applied Optimization

Volume 81

Series Editors:

Panos M. Pardalos
University of Florida, U.S.A.

Donald W. Hearn
University of Florida, U.S.A.

Applied Mathematical Modelling of Engineering Problems

By

Natali Hritonenko
Prairie View A&M University
Department of Mathematics
Prairie View, TX 77446
U.S.A.

Yuri Yatsenko
Houston Baptist University
College of Business and Economics
Houston, TX 77074
U.S.A.

Distributors for North, Central and South America:
Kluwer Academic Publishers
101 Philip Drive
Assinippi Park
Norwell, Massachusetts 02061 USA
Telephone (781) 871-6600
Fax (781) 871-6528
E-Mail <kluwer@wkap.com>

Distributors for all other countries:
Kluwer Academic Publishers Group
Post Office Box 322
3300 AH Dordrecht, THE NETHERLANDS
Telephone 31 78 6576 000
Fax 31 78 6576 474
E-Mail <orderdept@wkap.nl>

Electronic Services <http://www.wkap.nl>

Library of Congress Cataloging-in-Publication

CIP info or:

Title: Applied Mathematical Modelling of Engineering Problems
Author: Hritonenko, Yatsenko
ISBN: 1-4020-7484-0

Printed on acid-free paper.

Printed in United Kingdom by Biddles/IBT Global

Contents

Preface

Mathematical modelling is an effective tool widely used in science and engineering. Mathematical models and methods provide a rigorous, systematic, and quantitative description of various real-world (physical, chemical, biological) phenomena. They help to understand and analyse the essential qualitative features of these phenomena, to organize and process data, and to design and develop complex engineering systems. A good model can identify possible future trouble areas in an engineering activity and predict the side effects of recommended engineering solutions in the long run.

Mathematical modelling is often considered to be more an art than a science and depends heavily on the experience and knowledge of the researchers involved. It requires both a good understanding of the nature of the process and a familiarity with the available models and methods. A researcher needs to decide on the process to be modelled and the required level of abstraction and aggregation. In some practical cases, the phenomena under study are of very high complexity, and any mathematical description is just a modest approximation. Unavoidable simplifications and approximations made during the modelling process can greatly alter the predicted behaviour of the real-world phenomena. That is why applied mathematical modelling does not make sense without first defining the purpose of modelling. Before a model is developed, a specific existing problem should be described and the possible implications and benefits of the model need to be explored. The modelling goals have to be realistic but not oversimplified. In general, the level of complexity of mathematical models varies substantially for different applied problems.

There are no universal mathematical models and modelling methods in applied science and engineering. Numerous models of various types and complexity have been constructed to resolve specific practical problems. Some of these models appear to be appropriate for a wider area and, after thorough investigation and acceptance, contribute to the classical stock of mathematical modelling. Many mathematical models of engineering systems have become standard calculation techniques, for example, in the study of the strength of materials, heat flows, resource consumption, construction design, and other engineering activities. Sometimes the same model describes different real-life processes in various areas. For example, the evolutionary equation and renewal equation portray the population age structure and the equipment replacement in production systems, the diffusion equation describes individuals' migration, pollution propagation in air and water, and propagation of technological innovations, and so on. Finding analogies between mathematical models in different applied areas is always beneficial for the development of mathematical techniques and for all areas of modelling.

Many good monographs and texts in applied mathematical modelling have been written during the last decades. It is worthwhile to describe a place of the present book among them. The whole variety of books on mathematical modelling may be roughly divided into the following three categories.

First, there are *model-oriented* books where existing mathematical models are classified and listed by their categories (Kapur, 1988; Saaty and Joyce, 1981; Fowler, 1997; and others). These books are characterized by brevity, wideness, an encyclopedic style, and extended collections of modelling examples and applications.

The next type, *application-oriented* (or *process-oriented*) books, is represented by a larger number of publications, dedicated to different areas of engineering, for example, (Lin and Segel, 1974) for continuum mechanical systems, (Vierck, 1967) for oscillation processes, (Landau and Lifschitz, 1986) for elasticity theory, (Filippi et al, 1999) for acoustics, (Murray, 1989; Hoppensteadt and Peskin, 1992) for bioengineering, the series (Friedman, 1988-1998) for various industrial applications, and so on. The advantages of books in this category include exhaustive description of the engineering field under study, detailed explanation of particular models' structure and selected investigation methods. Such books concentrate on standard methods of analysis and show clear connections between mathematics and physical reality.

The last and the most common category consists of *method-oriented* books, with examples of (Fulford, Forrester, and Jones, 1997) for modeling with differential equations, (Corduneanu, 1991) for Volterra integral equations, (Sharkovsky, Maistrenko, and Romanenko, 1994) for difference equations, (Rubinov, 2000) for optimization methods, (Whitham, 1974) for wave theory, (Deans, 1983) for Radon transform, (Antonelli, Ingarden and Matsumoto, 1993) for the theory of sprays, and so on. These books mostly work within a selected field of mathematics. They are distinguished by in-depth analysis, an exhaustive explanation of a particular mathematical method or tool, its state-of-arts, potential, specific features and all known essential results.

The present monograph is a mixture of the above three categories and reflects the specific and unique experience of the authors. It is primarily model-oriented, then application-oriented, and finally method-oriented. The subject of the book is related to "know-how": how to construct specific applied models, how to adjust or modify them to a new engineering environment or more realistic assumptions, how to analyze models for the purpose of investigating real life phenomena, and how the models can extend our knowledge about a specific engineering process.

In the present book, the authors deal with mathematical models across a *wide range of practical applications*. They consider general modelling techniques, explain basic underlying physical laws, and transform them into a set of mathematical equations.

Particular attention is paid to the common features of the modelling process in various applied fields as well as to complications and generalizations of models. Engineering practice often drives investigation of new features of engineering systems, which requires either the development of new models or a substantial modification of known models (often, by using a new mathematical technique). The authors point out examples of slightly modified problems that lead to changes in the equation type. The mathematical nature of the derived equations is an important but secondary issue as compared with accounting for additional realistic assumptions.

From a *methodological standpoint*, the book is mainly related to continuous deterministic models, although some stochastic models are considered in the discussion of diffusion processes (Chapter 5). The models use various mathematical tools, including differential, difference and integral equations, the calculus of variations and optimal control theory. Special attention is paid to the authors' favorite mathematical tool – integral equations. Integral equations represent a more general tool than differential equations. The modification of an applied engineering problem often means a transformation from specific techniques based on differential equations to more general integral models. However, the authors are careful not to overestimate the importance of integral equations.

The range of applications covered in this monograph is diverse and includes mechanical, physical and electrical processes (lumped and continuous), environmental contamination, bioengineering and population control, production systems and technical equipment renovation. The authors want to show a variety of models and do not attempt to cover all possible engineering applications. They excluded from the book scope many interesting examples, such as nonlinear chemical engineering processes (Aiba et al, 1973; Bailey, 1991), and others. In contrast to their previous books (Hritonenko & Yatsenko, 1996; Hritonenko & Yatsenko, 1999), the monograph deals with local processes and systems rather than with large-scale and global ones.

All *chapters* of the monograph are independent and can be studied separately or in any sequence. However, there are many cross-references in the book because of the common nature of the subject.

The chapters are structured using the principle of moving from the simple to the complex. They start from a description of the model as constructed for rather simple generic and well-known (sometimes, classic) applied problems, often created under ideal conditions. When explaining how mathematical relations can describe basic physical processes, the authors often refer to the classical sources of mathematical physics (Courant & Hilbert, 1955; Morse and Feshbach, 1953) and modern application-oriented books (Svobodny, 1998; Logan, 1987; Kapur, 1988; Lin and Segel, 1974; and others). These basic examples are followed by less known and more advanced problems of an applied nature. Most of the advanced applied models discussed in the monograph reflect the authors' research experience. Some of them have been taken from specific engineering areas and may be new for the reader. The material is presented so as to minimize the required mathematical/ engineering preparation and encourage the further study of certain selected modelling topics.

There is a variety of relevant methods of analysis for various types of applied models. The authors concentrate on the way of a model construction and mention only important features and results obtained for the models. Some selected methods of analysis are briefly illustrated in each chapter of this book. The last two chapters contain more extensive description of two selected investigation techniques – bifurcation analysis and optimal control – applied to integral dynamic models with controlled memory. These techniques represent powerful methods of modern analysis. The authors were the first to develop and apply these methods to the integral models. The obtained results are new and important for applications.

The monograph can serve as a professional reference for mathematicians, engineers, applied or industrial scientists, and advanced students in mathematics or engineering. Furthermore, it provides excellent material for under graduate and graduate courses in mathematical modeling.

A brief description of the book content is given below.

Review of Book Content

Chapter 1 “SOME BASIC MODELS OF PHYSICAL SYSTEMS” has an introductory character. It discusses selected simple models of the dynamics of well-known physical (mechanical and electrical) processes that are of interest for numerous applied engineering problems. Later on in this book these and similar processes will be examined at a higher level of mathematical sophistication. The goal of this chapter is to refresh readers‘ knowledge and prepare them for more complex topics. Section 1 deals with particle mechanics, including projectile ballistics problems and linear and nonlinear oscillations. Section 2 is devoted to physical inverse problems, from Abel’s sliding particle to modern problems of computerized tomography. The chapter leads the reader through relatively elementary modelling of these phenomena, including some classical ordinary differential and integral equations and integral transforms.

Chapter 2 “MODELS OF CONTINUUM MECHANICAL SYSTEMS” is devoted to applied mathematical models of continuum media (solid, fluid and gas) that represent one of the most important techniques of applied mathematics. Such models are the source of the major applied partial differential equations such as the heat equation, Laplace and Poisson equations, diffusion equation, wave equation; and they reflect various phenomena like elasticity, viscosity, diffusion, dispersion, wave propagation, refraction, scattering and reflection. Developing models of continuum mechanics combines the universal conservation laws (balances of mass, momentum, and energy) that describe general physical phenomena with more detailed state equations, putting them into the form of a model. The conservation laws can be written in integral and differential forms. The integral form is more fundamental and is especially useful in cases when the differential form is not valid. In this chapter the authors provide basic conservation laws and briefly analyze the resulting models for one-dimensional and three-dimensional cases. A basic example is the model of the elastic motion of a thin bar, but there is also discussion of fluid and gas dynamics.

Deriving the motion equations does not yet provide a workable applied model. A researcher needs to decide which of a model's aspects are relevant and which are not, determine possible constraints and control influences, initial and boundary conditions, etc. To proceed further with the modelling process, more particular applied systems and modelling goals are required. To illustrate this point, in Section 4 the authors consider the applied models of water transport and contamination that are used for the control and conservation of water resources and flood prediction (in particular, for forecasting the radioactive contamination of Ukrainian rivers after the Chernobyl nuclear disaster). These models take into account major hydrophysical processes such as wind and water currents, water stratification, drift of suspended particles, drift of involved particles, the pollution transfer in solute and suspense, and the pollution accumulation in bottom sediments. A stationary flow problem is investigated for a two-dimensional model using stream function. Such problems arise in the prediction of spring floods.

Chapter 3 "VARIATIONAL MODELS AND STRUCTURAL STABILITY OF PHYSICAL SYSTEMS" deals with variational techniques applied to the dynamics of mechanical and physical systems and processes. In Chapter 2 the motion equations of continuum systems have been derived from conservation laws. Variational principles provide an alternative method of obtaining the motion equations based on the calculus of variations, a classical mathematical technique developed over two hundred years ago and applied to many engineering problems. In Section 1.1, one of the major variational principles, Hamilton's principle, is applied to several classic examples of continuum mechanics (string, bar, membrane and plate vibrations). The use of variational models in investigating spectral problems is illustrated in Section 1.2.

Variational techniques are also useful in formulating and investigating new models, especially if one is interested in aggregate structural parameters rather than a detailed description of the process's distributed characteristics. Some variational models for structural stability problems are discussed in

Section 2. Such problems consist of finding critical values of external parameters when a system is losing its structure or shape. First, the classic mechanical example of "a buckling rod" is considered. Next, two more specific and complex variational models from capillarity and plasticity areas are analyzed.

Chapter 4 "INTEGRAL MODELS OF PHYSICAL SYSTEMS" concentrates on integral equations used in applied modelling. Integral equations can describe global situations that cannot be modeled by differential equations. On the other hand, all models based on differential equations may be converted to integral equations. While general physical laws usually have an integral form, the derived equations of motion are often differential. Modifications of a mathematical model may require going back to an integral form. Possible examples include many problems of viscoelasticity, creep theory, superfluidity, aeroelasticity, coagulation and meteorology, electromagnetism, radiation transfer, radiophysics, electronic lithography, etc.

Section 1 describes two major techniques for obtaining integral models: conversion of differential models and a primary integral description of a process under study. Two powerful methods are considered there: Green's functions and the boundary integral equations method. The following sections deal with certain applied models in acoustic engineering and mining engineering. In Section 2, the method of boundary integral equations is applied to a problem of environmental noise propagation from road traffic (cars, train) over grass-covered ground. An integral model of a mine elevator considered in Section 3 describes the dynamics of mine rope as a one–dimensional continuum of variable length.

Chapter 5 "MODELLING IN BIOENGINEERING" is devoted to applied mathematical modelling in the rapidly growing field of bioengineering. There is a wide diversity of engineering problems related to this field. The impact of mathematical modelling methods in bioengineering has so far been limited, but this is now changing. Using biotechnology, industries can replaced chemical-based manufacturing processes with bioengineering processes, which are often less expensive, faster and safer for the environment. Genetic engineering can improve the by-products of microorganism-based industrial processes in order to make them commercially profitable. Biotechnology plays an increasing role in food production. Yeast makes bread rise; bacteria create yogurt; enzymes make cheese. We can now use genetic engineering to fabricate a desirable protein or enzyme. Moreover, these products or their by-products can be produced more effectively than from their natural source and on a larger industrial scale. In all these processes, the key is the natural reproduction of a population of some cells or microorganisms. While the relevant mathematical models have been thoroughly investigated, there still remain some interesting problems.

This chapter presents various models used in bioengineering applications, from the famous Malthus, Verhulst-Pearl, Fisher, and May population models to Markov processes and nonlinear models of population age-distribution. The mathematical tools include ordinary and partial differential equations, difference, integral equations and stochastic models. Important characteristics of the models (stationary points, stability, bifurcations) are discussed. In addition, the authors introduce a new nonlinear integral model of a population age-structure. The second part of this chapter is concerned with qualitative bifurcation analysis of this model. The analysis reveals oscillatory and stationary solutions of the model as well as the bifurcation values of parameters at which a new stationary solution appears. Connections between possible bifurcations in integral and difference models are established and discussed. The analysis shows the presence of new bifurcations and oscillatory regimes in the nonlinear integral model. Certain residual problems are discussed.

Chapter 6 “MODELLING OF TECHNOLOGICAL RENOVATION IN PRODUCTION SYSTEMS” concentrates on modelling the rational replacement (renovation) of industrial equipment under conditions of technological change. Section 1 briefly examines traditional approaches for the modelling of technological restoration in production systems, such as aggregated models of optimal investments, partial differential models of equipment age structure; and statistical models of equipment replacement (the renewal equation). Modern economic growth is characterized by structural changes based on the introduction of new technologies into equipment. The replacement and renovation of the technologies in industrial environments undergoing technical change is one of the key aspects of economic development. However, the modelling and control of the technologies’ lifetime has received rather little attention in contrast to other aspects of industrial modelling. It requires developing special models that consider the service interval (lifetime) and age structure of the equipment.

From a mathematical viewpoint, this chapter is devoted to the optimal control of dynamical systems governed by specific non-linear Volterra integral equations with unknown functions in the limits of integration. In Section 2 aggregated and disaggregated integral models of production systems with endogenous equipment lifetime are constructed, discussed and compared with other explorations. Optimization problems in these models are investigated in Section 3. Their asymptotic analysis reveals interesting insights into general patterns for of the rational lifetime of capital equipment in production systems under technological change. Under certain simplified assumptions, so-called *turnpike properties* are established for the optimal lifetime of equipment by the authors, for the first time. Such properties are well known for other (non-integral) applied models of mathematical economics and describe general tendencies of economic dynamics. Section 3 presents discussion of the obtained results and some open problems. Detailed mathematical reasoning and proofs are contained in Section 4.

Chapter 7 "APPENDIX" contains miscellaneous definitions and results of the calculus, functional analysis, mathematical models, and differential and integral equations that are used in the monograph. It contains a basic classification of mathematical models that matches the properties of different models with the features of real-life systems.

Acknowledgements

In conclusion, we wish to express our sincere gratitude to Professor Efim Galperin from the University of Quebec in Montreal (Canada), Professor Gerald Silverberg from the Maastricht Economic Research Institute of Innovation and Technology (the Netherlands), Professor Alex Rubinov from the University of Ballarat (Australia), Doctor Klauss Puhlmann from Grande Yellowhead Regional Division No.35 (Canada), Professor Arkadii Chikrii from the Cybernetics Institute of the Ukrainian Academy of Sciences (Kiev, Ukraine), Professor Rafail Gabasov from the Byelorussian University (Minsk, Belarus), and Marsha Shelton from Prairie View A&M University for their interest and moral support of our research.

We are grateful to Professor Michael Bourke from Houston Baptist University and Professor Orion Ciftja from Prairie View A&M University for useful remarks and improvements in certain sections. Also, we wish to thank the Office of the Vice-President for Research and Development of Prairie View A&M University for a support grant that has greatly assisted us in the final stages of manuscript preparation.

Natali Hritonenko

Yuri Yatsenko

Chapter 1

Some Basic Models Of Physical Systems

This chapter presents a collection of selected models of basic physical (mechanical and electrical) processes that have been routinely used in solving numerous applied engineering problems. While there are many well-known established areas appropriate for using as an introduction into applied mathematical modelling, the authors have chosen the following two topics :

a) the mechanics of a material particle, including projectile ballistics, linear and nonlinear oscillations (Section 1),

b) physical inverse problems, starting from Abel's sliding particle to modern problems of computerized tomography (Section 2).

Most of the examples that are considered belong to classical topics. They have an elaborated analytic theory and well-developed computer simulation methods. Mathematical tools used in this chapter include ordinary differential and integral equations. The authors do not attempt to cover all known results for the models (which is obviously impossible). Rather the goal of this chapter is to explain how a mathematical model may be constructed for a physical process. Only the models' most prominent features and methods are mentioned.

1. BASIC MODELS OF PARTICLE DYNAMICS

In this section, the dynamics of *finite-dimensional mechanical systems* is investigated. These systems consist of a combination of separate (interconnected) material particles and can be described by a finite number of scalar mechanical parameters (mass, force, momentum, position, velocity, acceleration, etc.). The parameters can depend on time (dynamic parameters) or they can be constant (static parameters). Time is the only independent variable in these models.

Finite-dimensional systems are often called *systems with lumped parameters* or *lumped systems*. The corresponding mathematical models of such systems are referred to as *lumped models* and described by *ordinary differential equations* (*ODE*) or one-dimensional *integral equations* (*IE*).

Another type of model is the *distributed model*, which includes more than one independent variable. The additional variables usually describe a position in space, meaning that the parameters of the system depend on time and space coordinates simultaneously. Such systems are called *continuum systems*, *infinite-dimensional systems*, *systems with distributed parameters,* or *distributed systems*. The distributed models are described by *partial differential equations* (*PDE*) or multi-dimensional integral equations. They represent more complex mathematical tools than lumped models. Certain applied systems may be distributed by their nature, for example, a river, an atmosphere, a bridge, an electric power transmission line, etc. The distributed models of continuum mechanics are investigated in Chapters 2-4, of bioengineering - in Chapter 5, and of technological renovation problems - in Chapter 6.

If at least one independent variable in a distributed model is a space coordinate, then it may be referred to as a *spatial variable*, and the model is a *spatial model.* In other cases, the additional independent variable in *distributed models* represents just another time (for example, the time of introducing new equipment into a production system) in addition to the current time (see Chapter 6).

In conclusion, the authors want to note that the above classification of mathematical models depends on the nature of a physical system as well as on the level of physical abstraction accepted during a modelling process. For example, one can investigate the oscillation of a mass on a string as a lumped problem (which is done in Section 1.2); on the other hand, the longitudinal waves in the string itself can be analysed as a distributed problem (see Section 4 of Chapter 2).

1.1 Motion of a Particle in Gravitational Field

The motion of a material particle in the gravitational field is a classical example of mathematical modelling, and it is probably the most used one during last centuries. It has been covered in many textbooks.

Let us consider a particle in a three-dimensional space and assume that the only forces acting on the particle are the gravity **G** of the earth and the medium (air) resistance **F**. The behavior of the particle is governed by the following two basic mechanisms (Lin and Segel, 1974; Dreyer, 1993; and others):

Newton's second law of motion: *The change of motion is proportional to the motive force impressed, and it takes place along the right line in which that force is impressed.*

Newton's law of universal gravitation: *The magnitude of the force* **G** *acting between two bodies of a point mass is proportional to their masses M and m and is inverse proportional to the square of the distance* **r** *between the bodies, i.e.*

$$G = \gamma Mm / r^2 , \tag{1.1}$$

where γ is the so-called *universal gravitation constant.*

Therefore, the general model of the particle motion is described by the following differential equation:

$$d(m\mathbf{v})/dt = \mathbf{F}(\mathbf{v}) + \mathbf{G}(\mathbf{x}), \qquad d\mathbf{x}/dt = \mathbf{v}, \tag{1.2}$$

where m is the mass of the particle,
$\mathbf{x} \in \Re^3$ is its position in space,
$\mathbf{v} \in \Re^3$ is its velocity.

In this case the motion is represented by the linear momentum $m\mathbf{v}$ of the particle. The gravitation force **G** depends on **x** because of the fundamental physical law (1.1). The resistance force **F** is determined by more specific physical relationships that depend on the medium. It is shown experimentally that the air resistance force **F** depends on the speed **v** and the magnitude of **F(v)** is directly proportional to **v** for low speeds ($\mathbf{v} < 25$ m/sec), proportional to the square of **v** for higher speeds (25 m/sec $< \mathbf{v} <$ 300 m/sec), increases drastically near the *sound barrier* ($\mathbf{v} \approx 335$ m/sec), and is proportional to **v** again for *supersonic speeds* ($\mathbf{v} > 340$ m/sec).

Three examples in the following subsections describe model (1.2) in more specific cases.

1.1.1 Vertical Projectile Problem.

Consider a body of mass m that is radially projected upward from the Earth's surface with an initial speed V_0. Let R denote the radius of the Earth and $y(t)$ denote the sought-for radial distance from the Earth's surface at time t. If one neglects the air resistance, the model of the body's dynamics (1.2) is described by the following *nonlinear differential equation* of the second order:

$$\frac{d^2 y}{dt^2} = -\frac{gR^2}{(y+R)^2}, \tag{1.3}$$

with the initial conditions :

$$y(0) = 0, \qquad dy/dt\,(0) = V_0, \tag{1.4}$$

The constant $g = \gamma M/R^2$ is called the *gravitation acceleration* on the Earth's surface (at y=0) where M is the Earth's mass (see (1.1)).

If the displacement y is small compared to R (it is true when the initial speed V_0 is not too large), then it leads to the simplified linear problem:

$$d^2y/dt^2 = -g, \quad y(0) = 0, \quad dy/dt(0) = V_0,$$

with the well-known elementary solution: $y(t) = -gt^2/2 + V_0 t$.

1.1.2 Free Fall with Air Resistance

In this case, the air resistance is considered but changing gravitational attraction is not taken into account. Let us consider a free fall with the initial speed of zero. In this case, there is no horizontal component in the velocity $\mathbf{v}$ and in the displacement $\mathbf{x}$, and the only motion is the vertical one. Let us choose the y-axis as the vertical axis upwards. Then equations (1.2) lead to

$$d^2y/dt^2 = -g - f(v), \tag{1.5}$$

$$y(0) = y_0, \qquad dy/dt(0) = 0,$$

or to the nonlinear differential equation of the first order

$$dv/dt = -g - f(v)\,, \quad v(0) = 0, \tag{1.6}$$

with respect to the speed v.

Whether the air resistance force $f(v)$ is suggested to be proportional to the speed v or to the to the square of v, the solution $v(t)$ of (1.6) tends to a constant speed V_∞, called the *terminal speed*, when $t \to \infty$.

In the case $f(v) = kv^2$, the terminal speed is $V_\infty = (g/k)^{1/2}$ and the (1.6) solution is

$$v(t) = -V_\infty \tanh\,(gt/V_\infty).$$

1.1.3 Plane Projectile Problem

Let us consider the motion of a projectile that is launched to an acute angle θ to the horizontal, so that the motion can be described in a plane. Choose the Cartesian coordinates with the origin at the point of launching the projectile, the x-axis in the horizontal direction of the initial speed, and the y-axis in the vertically upward.

We restrict ourselves with the case of a moderate initial speed V_0 (not the rocket case), so the air resistance is directly proportional to the speed (with a coefficient k) and the gravitational attraction is constant.

Then the model (1.2) is described by the following linear differential equations:

$$\begin{aligned} dv_x/dt &= -kv_x/m\,, & v_x(0) &= V_0\cos\theta, \\ dv_y/dt &= -g - kv_y/m\,, & v_y(0) &= V_0\sin\theta, \end{aligned} \tag{1.7}$$

The solution is

$$v_x = V_0 \cos\theta\, e^{-kt/m},$$

$$v_y = V_0 \sin\theta\, e^{-kt/m} - mg(1 - e^{-kt/m})/k.$$

The case when the air resistance is proportional to the square of the speed V_0 is considered in (Lambe and Tranter, 1961).

1.1.4 More General Ballistic Problems

Despite their classical nature, the models described above are of applied character and have been used in practice. Examples include calculating the ballistics of gun shells, intercontinental missiles, spacecrafts, etc. In such cases, general model (1.2) of the particle motion takes some of the following additional physical phenomena and effects into account (Kapur, 1990):

- rotation of the Earth produces additional acceleration of the particle;
- the air resistance force might be related to the speed in more complicated manner. It is proportional to different powers for different ranges of the speed (see the note at the beginning of Section 1.1);
- some additional force might apply short time after the start while a missile uses its fuel and the mass of the missile is changed because of the consumption of the fuel;
- changing wind velocity, air humidity and air pressure impact the motion of the particle;
- since a real-life object is not a particle but a rigid body, its motion about its center of gravity has to be considered. For example, when a gun shell comes out of the gun, it is rotating for better stability of the motion;
- for high speeds, the shape of the object determines its aerodynamics and heating during the flight.

Naturally, such models cannot be resolved in elementary functions and they have special methods of their analysis (analytical and numeric). An interested reader may find more details in specific mechanical textbooks and monographs.

1.2 One-Dimensional Mechanical Vibrations

Vibrating motion (oscillation) represents the heart of many engineering systems and processes. In some cases, the vibrations provide a negative impact onto the system and should be avoided; in others, they deliver required operations and need to be as large and efficient as possible (Lin and Segel, 1974; Logan, 1987; Vierck, 1967, etc.).

Compared with the previous section, models of vibrating motion include a new type of forces (restoring elastic force) that depend on the position of the material particle. Then Newton's second law of inertial motion leads to the models of motion described by the ordinary differential equations of the second order.

1.2.1 Linear Oscillator

In this section, we consider linear oscillations of a material particle with respect to a stable static equilibrium position (the center of oscillations). Since the equilibrium is stable, it implies the presence of some restoring force acting upon the particle in order to restore the equilibrium.

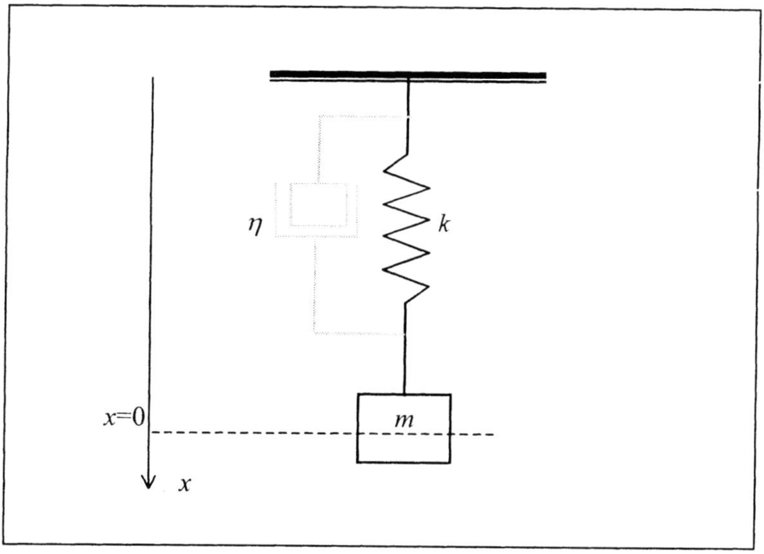

Figure 1.1. The mass-spring system is an example of the linear oscillator (1.8); the grayed part corresponds to an optional dashpot mechanism that provides viscous damping and corresponds to the equation (1.11) (spring-mass-dashpot system).

The simplest mechanical system of such kind is the *spring-mass* system (see Figure 1.1) that consists of a mass m and a (massless) elastic spring that connects the mass to a motionless rigid boundary AB. Let the coordinate axis x be parallel to the string and the point $x=0$ coincide with the position of stable static equilibrium of the mass.

The motion of the spring-mass system is again governed by the Newton's second law $d(m\mathbf{v})/dt=\mathbf{F}$ of inertial motion (see Section 1.1) that leads to the model of the mass oscillation described by the following linear differential equation of the second order:

$$md^2x/dt^2 = -kx. \tag{1.8}$$

Oscillations in real-life systems are usually deteriorated without external forces applied because of various friction and damping phenomena. The simplest model of damping is the *viscous damping* that is proportional to the particle velocity:

$$F_{el} = \eta dx/dt.$$

where η is the *viscosity coefficient.*

Then the mass oscillations are described by the following linear differential equation:

$$m\,d^2x/dt^2 + \eta\,dx/dt + kx = 0. \tag{1.9}$$

The spring-mass system (1.9) with viscous damping is known as the *spring-mass-dashpot* system. The equation (1.9) is resolved in elementary functions, and the major dynamics is represented by the decaying oscillations if the damping parameter $\lambda=\eta/2m$ is not too large, $0<\lambda<\omega_0$,

$$x(t) = A\exp(-\lambda t)\cos(\omega t+\phi)$$

with the effective frequency ω and the amplitude A. The effective frequency is connected with the *natural frequency* (*frequency of undamped vibrations*) $\omega_0 = (k/m)^{1/2}$ as $\omega^2 = \omega_0^{\,2} - \lambda^2$.

Another important physical oscillator is an *inductive electrical circuit* (*RLC circuit*) that represents a serial connection of three basic electric elements: an inductivity L, a capacitor C and a resistor R. The dynamics of the circuit is governed by the Kirchoff-Ohm laws and is described by the following linear differential equation with respect to the electrical current $i(t)$:

$$L\, d^2i/dt^2 + R\, di/dt + i/C = 0. \tag{1.10}$$

This equation (1.10) is identical to the equation (1.9). We will return to this important *electrical-mechanical analogy* later in this chapter.

1.2.2 Forced Linear Vibrations and Resonance.

If a spring-mass-dashpot system is subjected to an external force F, its equation becomes

$$m\, d^2x/dt^2 + \eta\, dx/dt + kx = F. \tag{1.11}$$

The linear differential equation (1.11) may be solved analytically for an arbitrary function $F(t)$ by the *variation of parameters formula*. However, the most important and interesting case for applications is the case of a *steady periodic force* $F(t)=F(t+T)$, T=const>0. This case produces the *resonance phenomenon,* which is widely used (or avoided) in applied engineering systems.

To illustrate the resonance effect, let us consider forced vibrations of the spring-mass-dashpot system (1.11) under the *harmonic excitation force* $F(t)=F_0\exp(i\omega t)$. Then the steady-state component (that does not decay in time) of the general solution of the equation (1.11) also has a harmonic form $x(t)=A\exp(i\omega t)$ with the same frequency ω and the amplitude A:

$$|A| = \frac{F_0 / m}{\sqrt{(\omega_0{}^2 - \omega^2)^2 + 4\lambda^2\omega^2}}, \quad \lambda = \frac{\eta}{2m}, \quad \omega_0 = \sqrt{\frac{k}{m}}. \tag{1.12}$$

It is easy to see that at ω close to ω_0, the amplitude A of the forced vibrations increases sharply and is determined only by the damping parameter λ. The phase portrait (1.12) on the (A,ω)-plane is shown on the Figure 1.2 and is described by the following "resonance peak"

$$\omega - \omega_0 = \pm 2\lambda \sqrt{\frac{F_0^{\ 2}}{4\lambda^2 A^2} - 1}.$$

If there is no damping (λ=0), then $A \to \infty$ as $\omega \to \omega_0$, which means that the forced vibrations indefinitely increase in time (the dotted lines on Figure 1.2). This type of the oscillating behavior is called *the resonance*. The resonance phenomenon plays overwhelming role in engineering practice. It creates a basis for necessary efficiency of many production processes, however, if handled wrongly, it often leads to destruction of a mechanical system.

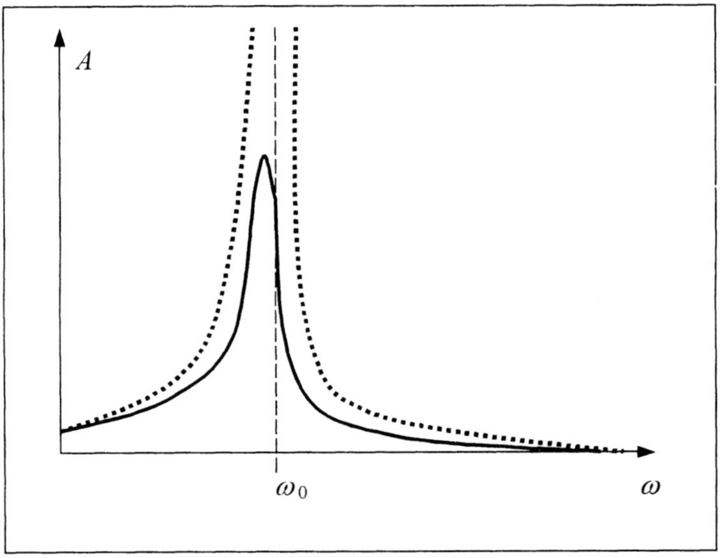

Figure 1.2. The resonance characteristic (1.12) of the linear oscillator (1.11) (the dotted lines correspond to the absence of damping λ=0).

1.2.3 Nonlinear Oscillators

The linear oscillator is a convenient theoretical abstraction that has appeared to be a very useful approximation for many applied scientific and engineering processes. However, real-life processes are almost always nonlinear. The sources of nonlinearity are different. They include nonlinear physical laws and external forces, nonlinear combinations of interacting linear systems, the scale of system motion, and so on.

One of the major research problems in applied modelling is the choice of a linear or nonlinear model. This choice need to be justified with respect to modelling goals, the impact of nonlinearity, and the complexity of simulation.

There are many nonlinear physical mechanisms in mechanical oscillations, which produce strange and complicated dynamics of the systems. To illustrate how nonlinear effects appear, different types of nonlinearities are discussed below.

1.2.3.1 Simple Pendulum

A simple pendulum represents a classical example of a *nonlinear* mechanical system. The pendulum consists of a bob of the mass m at the end of an unflexible massless rod of the length l connected to a motionless rigid boundary.

The dynamics of the pendulum is completely described by the angle θ between the rod and the vertical. Let us consider $\theta=\theta(t)$ as the sought-for function of time. If there is no external restriction on the pendulum motion, it has a stable equilibrium $\theta=0$ in the lowest position (when the rod is vertical). The only external force is the downward gravity force **G.** The component of **G** parallel to the direction of the bob motion has the magnitude equal to $mg\sin\theta$. (i.e., a nonlinear function of θ). This force is the only restoring force for the bob when it deviates from the equilibrium position $\theta=0$.

Using Newton's second law $d(m\mathbf{v})/dt=\mathbf{F}$ (see Section 1.1) and the formula $|\mathbf{v}|=l\theta$ for the linear velocity of the bob, we derive a model of the pendulum motion as the following *nonlinear differential equation* of the second order:

$$d^2\theta/dt^2 + (g/l)\sin\theta = 0. \tag{1.13}$$

So, the pendulum oscillation represents a natural nonlinear physical process. In the case of small angle displacements θ, the approximate expression $\sin\theta = \theta + o(\theta^2)$ holds, and the equation (1.13) of pendulum motion becomes a linear oscillator (1.8):

$$d^2\theta/dt^2 + (g/l)\theta = 0 \qquad (1.14)$$

with the natural frequency $\omega_0 = (g/l)^{1/2}$. So, at least for small initial angles, the pendulum bob oscillates around the lowest equilibrium position $\theta=0$. For large enough values of the initial angles and angular velocities, the bob will spin around the pendulum support.

1.2.3.2 Nonlinear Restoring Mechanical Forces

Let us consider the spring-mass-dashpot system from Section 1.2.1 again. Hook's law about direct proportionality of the deformation and the corresponding force represents a constitutive relation (see also Chapter 2) that is valid only for small deformation values.

For large enough displacements, the elastic restoring force of the spring becomes a nonlinear function of the displacement: $F_{el}=F_{el}(x)$, $F_{el}'(x) \neq \text{const}$. For the majority of materials, including steel, the elastic restoring force will increase for larger displacements. Such strings are known as "*hard strings*". For some plastic materials, the stiffness decreases when displacements become larger (a "*soft string*"). This effect is usually modelled by the following nonlinear expression:

$$F_{el} = kx + \kappa x^3, \qquad (1.15)$$

where the parameter k is the same as in the linear model (1.8) and the parameter κ characterizes the nonlinearity; $\kappa>0$ for "hard strings" and $\kappa<0$ for "soft strings". The above cubic power function κx^3 is chosen because normally the restoring force is the same regardless whether the material is stressed or elongated, i.e. $F_{el}(-x) = -F_{el}(x)$. However, some very specific plastic materials soften as they are stretched and stiffen as they are compressed. Then the expression (1.15) for $F_{el}(x)$ should include a quadratic term as well.

Using Newton's second law, we obtain the following nonlinear differential equation of the spring-mass-dashpot system with a nonlinear string:

$$m\, d^2x/dt^2 + \eta\, dx/dt + kx + \kappa x^3 = F. \tag{1.16}$$

The equation (1.16) is investigated below in Section 1.2.4. It is easy to see that the pendulum motion (1.13) corresponds to a "soft string" case.

Other possible generalizations of Hook's law lead to even more complicated mathematical models (see Section 2.5 of Chapter 2).

Nonlinear laws often describe the components of *electrical circuits*, too. First of all, a variety of nonlinear electrical components have been specifically designed to implement some nonlinear effects. For example, a diode or triode (semiconductor or vacuum-tube) has a significantly nonlinear current-voltage (I-V) characteristic curve $I=f(V)$, where $f(V)$ may increase for some diapasons of V and decrease for others. Correspondingly, any electric circuit with a diode or triode is described by a nonlinear equation.

For other components, the nonlinearity is undesirable but an unavoidable feature caused by the physics of a process. Thus, the electromagnetic flux ϕ is normally proportional to the electrical current I in vacuum but not in steel coils because of the *saturation effect* inherent for the electromagnetic processes in iron. The corresponding modelling expression $I=\Phi(\phi)$ appears to be similar to the "hard string" restoring force.

1.2.3.3 Nonlinear Resistance Forces (Damping and Friction)

The resistance forces in mechanical motion are often nonlinear depending on the medium and the speed of motion. The resistance forces include : (a) damping and (b) friction.

The *viscous damping force* was already discussed in Sections 1.1 and 1.2.1. This force is linearly dependent on the speed of motion for low values of the speed (more exactly for low values of Reynolds number). It is proportional to the square of the speed in the case of higher speeds. For example, if a pendulum or another oscillator is considered in the air medium, then the linear aerodynamic damping dependence is valid only in the case of small initial deviations and speeds.

The *damping force* may also depend on the position of the oscillating particle. An example is given by a *Van der Pol oscillator* equation

$$m\, d^2x/dt^2 + (\eta_1 x^2 - \eta_0)\, dx/dt + kx = 0, \tag{1.17}$$

that appears in self-exciting electrical circuits with active elements.

The *friction force* is a resistance force that arises when a moving mass (a particle) has a contact with another rigid surface. This force may be of two different types depending on the surface properties: *wet friction* and *dry friction*. The wet friction force behaves similarly to the viscous damping.

The *dry friction* force is always sufficiently nonlinear and consists of two components:

a) *sticking friction*, which is the minimum force that must be applied in order for the motion to begin;

b) *dynamic friction*, which arises during the motion and may depend on the speed, acceleration and position of the moving particle.

The oscillations of a particle subjected to the dry friction have a non-harmonic relaxation character even for harmonic excitation forces.

1.2.4 Nonlinear Vibrations and Resonance.

From the theoretical point of view, the difference between linear and nonlinear oscillators is significant. While the linear oscillator model is resolved in elementary functions, a nonlinear oscillator dynamics can be usually investigated using approximate methods and possess very specific and odd features in special cases. The superposition principle is not valid for nonlinear equations, so their solutions cannot be combined as sums of individual components. The major nonlinear effects include:

- The output oscillations are not of sinusoidal form for a sinusoidal input (excitation force).
- The oscillation form depends on the amplitude of the input.
- The oscillations may contain *subharmonic* and *superharmonic* components with the frequencies $\omega/2$, $\omega/3, \ldots$, 2ω, 3ω, ... for a pure sinusoidal input of the frequency ω.
- In general, the output dynamics may contain frequencies that are absent in the input.
- The resonance frequency depends on the amplitude of oscillations; the amplitude of oscillation may jump when the frequency of external periodic force passes the resonance domain.

– The *hysteresis effect* may appear when the same system behaves differently depending on whether the frequency of external periodic force is increased or decreased.

In practice, nonlinearity is often small in terms of its impact and the oscillator dynamics is close in some sense to a linear case. There is a set of classic approximate methods (*small parameter methods, averaging methods, perturbation methods,* etc.) developed for such cases (Poincare, 1993; Bogoliubov and Mitropolsky, 1961; Nayfeh and Mook, 1979; and others). They produce a solution of a nonlinear model as an expansion with respect to a small nonlinear parameter in the model equation (in doing so, the 0-approximation usually corresponds to the linear case).

We illustrate some of these properties for the nonlinear model (1.16) of spring-mass-dashpot system with a "lightly nonlinear" string. To separate a small parameter in the equation (1.16), one can rewrite it in the following *"normalized" form*:

$$d^2x/dt^2 + 2\lambda\, dx/dt + x + \beta x^3 = F, \tag{1.18}$$

where $\beta>0$ for "hard strings" and $\beta<0$ for "soft strings". Now let us assume that $|\beta|<<1$ (the string nonlinearity is small) and $\lambda<<1$ (the damping is small).

Even under these simplified assumptions, the nonlinear differential equation (1.18) cannot be solved analytically. Approximate methods for this equation take back to Poincare's works. One of the major results obtained using various techniques and confirmed by numerous experiments consists of the nonlinear dependence of the resonance frequency ω on the amplitude A of oscillations. Namely in the absence of damping ($\lambda=0$):

$$\omega^2 = 1 + \frac{3}{4}\beta A^2 \tag{1.19}$$

(the dashed parabola in Figure 1.3).

For the case of the small string nonlinearities $\beta<<1$ and small damping $\lambda<<1$, the following approximate resonance curve is obtained:

$$\omega - 1 = \frac{3}{4}\beta A^2 \pm 2\lambda\sqrt{\frac{C^2}{4\lambda^2 A^2} - 1}. \tag{1.20}$$

The function (1.20) for hard strings (β>0) is shown on the Figure 1.3 with the solid line. The nonlinear resonance is characterized with a resonance peak that is tilted with respect to the x-axis by a sloping that depends on the nonlinearity parameter β (compare with the resonance peak for the linear oscillator (1.11) in the Figure 1.2). The relationship (1.20) is multi-valued for some diapason of ω and explains the nonlinear effects (including *hysteresis*) mentioned above at the beginning of the section.

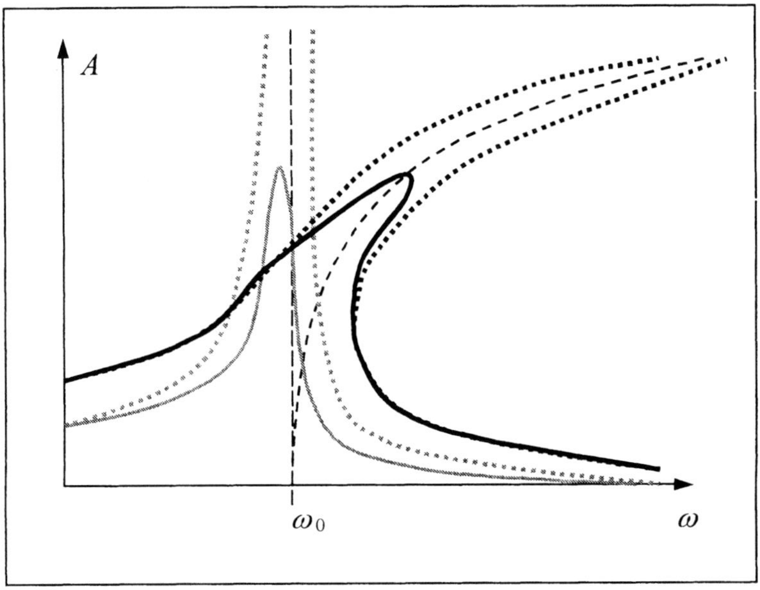

Figure 1.3. The resonance characteristics (1.20) of the nonlinear oscillator (1.18) (the dotted lines correspond to the absence of damping at λ=0; the grayed lines correspond to the linear case (1.11) at β=0).

1.2.5 Nonlinear Electrical-Mechanical Systems

Engineering devices often include a combination of several linear dynamic systems that interact in a nonlinear manner. There is a wide variety of such devices in electric engineering, ranging from a simple electrical shaver to powerful electrical vibrators for road construction. The key function of such devices lies in the transformation of the electrical power into a mechanical action. This transformation is often more efficient to be implemented in a nonlinear way that depends on specifics of the device.

Let us consider an electrical-mechanical system that consists of a mechanical spring-mass-dashpot oscillator with linear forces and electrical excitation. The electrical *R-L* circuit of the device has a coil with a movable core, and the coil core represents the oscillating mass (see Figure 1.4). If an external periodic voltage U applies to the circuit, then the corresponding electromagnetic force F_{em} is developed in the coil and excites the oscillations of the mass (the coil core).

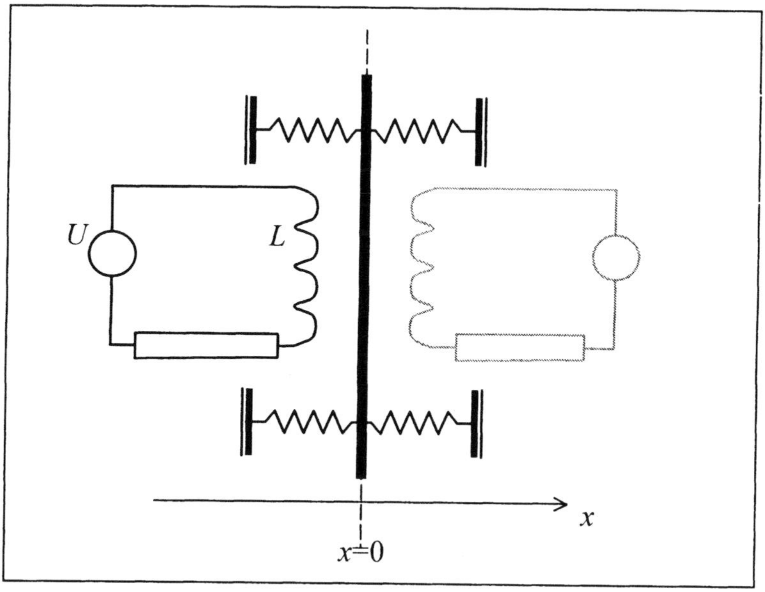

Figure 1.4. Dynamics of the spring-mass-dashpot oscillator with periodic electromagnetic excitation is described by the equations (1.21)-(1.22); the grayed part represents the second optional electrical *RL* circuit that provides more even electromagnetic force (see equations (1.23)).

The corresponding mathematical model of the electrical-mechanical system under study includes two equations:

- the equation for the mechanical oscillation :

$$m\, d^2x/dt^2 + \eta\ dx/dt + kx = F_{em}, \tag{1.21}$$

- the equation for the current i through the R-L circuit :

$$L\, di/dt + R\, i = U. \tag{1.22}$$

Even if we omit all possible internal nonlinearities of the mechanical restoring and damping forces and of the electromagnetic flux in iron coil (the *saturation effect*), the above electrical-mechanical system is still nonlinear because of its construction.

There are two sources of nonlinearity in the described simple device:

a) the electromagnetic force developed in the coil is proportional to the square of the electromagnetic flux ϕ :

$$F_{em}=a\phi^2;$$

b) the inductivity L of the electrical circuit depends on the coil air gap that is a function of the sought-for mass displacement x :

$$L=L_0/(1+x).$$

Let us assume the characteristic curve $i=\Phi(\phi)$ of the coil to be linear, i.e $i=b\phi$, then the mathematical model (1.21)-(1.22) has the form:

$$m\, d^2x/dt^2 + \eta\, dx/dt + kx = (a/b^2)\, i^2,$$

$$L_0/(1+x)\, di/dt + R\, i = U.$$

In general, a nonlinearity leads to non-harmonic relaxation-type vibrations of an electrical-mechanical system even for a harmonic excitation and, as result, to higher energy losses than in a linear case.

Various modifications of the above simple system were developed in attempt to provide more efficient ("less non-linear") functionality required for powerful vibration devices. One of the modifications is adding the second symmetrical electrical R-L circuit on the opposite side of the oscillating mass, which is powered by external voltage $-U$. Then, the total electromagnetic force in the coil is equal to the difference of two forces

$$F_{em}= F_{em1} - F_{em2} = a(\phi_1{}^2 - \phi_2{}^2) = a(i_1{}^2 - i_2{}^2)/b^2 = a(i_1 - i_2)(i_1 + i_2)/b^2$$

and the dynamics of the system becomes more smooth and efficient.

Then, the mathematical model (1.21)-(1.22) consists of the mechanical vibration equation and two equations for the currents i_1 and i_2 in each circuit:

$$\begin{aligned} & m\, d^2x/dt^2 + \eta\, dx/dt + kx = (a/b^2)(i_1 - i_2)(i_1 + i_2), \\ & L_0/(1+x)\, di_1/dt + R\, i_1 = U, \\ & L_0/(1-x)\, di_2/dt + R\, i_2 = -U. \end{aligned} \tag{1.23}$$

A preliminary analysis of the nonlinear system (1.23) shows that, because of the special symmetry of both electrical circuits, their currents i_1 and i_2 are characterized by the following relationship: $(i_1+i_2) \approx$ const. This means that the total electromagnetic force F_{em} is approximately harmonic for a harmonic external voltage U and harmonic currents i_1 and i_2.

However the electrical-mechanical system is still nonlinear due to the nonlinear inductivities $L_0/(1\pm x)$ and requires special methods of analysis.

2. INVERSE PROBLEMS AND INTEGRAL MODELS

Determination of the dynamics of a mechanical (physical) system under the given forces is commonly classified as a *direct problem*. *Inverse problems* are concerned with finding out some initial geometric or physical characteristics of the system for given dynamics. Inverse problems play an important role in numerous engineering applications and often lead to integral equations (Ramm, 1980; Ramm and Katsevich, 1996).

This section presents crtain basic inverse problems for mechanical and physical systems. The first problem (a sliding particle) belongs to the field of particle mechanics and leads to the first and the most famous integral equation – *Abel's equation*. The second problem describes a similar mechanical example (a sliding chain) and is adapted from (Corduneanu,

1991). Next, the important physical inverse problems of computerized tomography are considered. They use Fourier and Radon integral transforms. All problems of this section are described by integral equations. Some other applied integral models of mechanical systems are illustrated in Chapters 2 and 4.

2.1 Sliding Particle and Abel's Equation

Let us consider a material particle of mass m that is sliding down along a smooth curve in a vertical plane, without friction, under the gravity force only. This mechanical system is described in many sources, for example, in (Porter, 1990; Corduneanu, 1991).

We choose a coordinate system Oxy in the vertical plane where the motion takes place as shown at Figure 1.5. The particle starts with zero initial velocity at the point (X,Y) and reaches the origin O at the time which is naturally a function of the initial height Y.

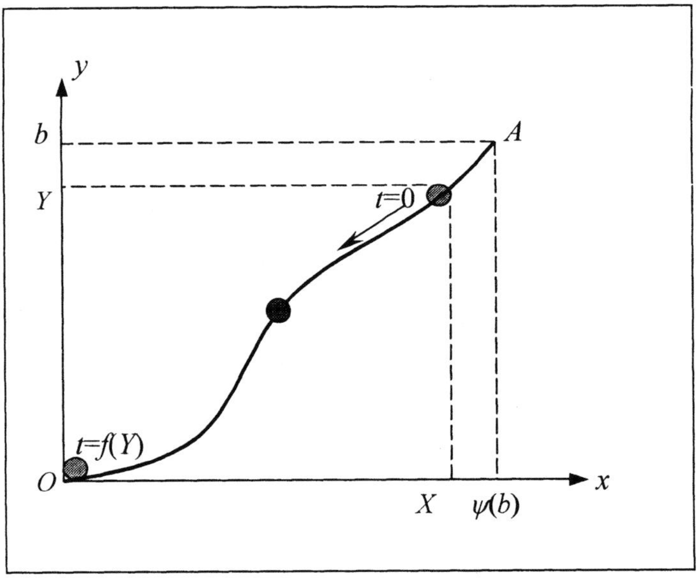

Figure 1.5. Dynamics of the particle M sliding down along the curve OA is described by Abel's integral equation (1.26)

To describe the motion, let us introduce two key functions:

- the time $f(Y)$ of descent from the height Y, and
- the shape $x=\psi(y)$ of the curve OA (along which the motion takes place).

The function $f(y)$ is defined by the dynamics of the motion whereas the function $\psi(y)$ is a geometric characteristics of the problem under study. Determination of the sought-for $f(y)$ under the given $\psi(y)$, $0\leq y\leq b$, is considered as a *direct problem* and finding $\psi(y)$ under the given $f(y)$, $0\leq y\leq b$, is considered as an *inverse problem.*

Let $\psi(.)$ be a smooth function and $\psi(0)=0$. Then the length s of the arc from the origin O to the generic point (x,y) is described by the following relationship:

$$ds/dy = \{1+ [\psi'(y)]^2\}^{1/2}. \tag{1.24}$$

As opposed to previous examples of this chapter, we the energy balance is used instead of Newton's laws to derive the equation of motion. One can find more information about relations between Newton's laws and energy conservation balances in Section 1 of Chapter 2.

The ***energy conservation law*** states that, *if all forces acting on a mechanical system are conservative (like the gravity force), then the total energy of the system is not changed.*

The total energy, $T+U$, of the particle M consists of its *potential energy* $U=mgy$ and the *kinetic energy* $T=\frac{1}{2}mv^2$, where v is the speed of the particle. Considering the total energy $T+U$ for $t=0$ and a generic time t, the equality $mgY = mgy + \frac{1}{2}mv^2$ is obtained Hence, the speed v is equal to $[2g(Y-y)]^{1/2}$. Using (1.24) and the obvious relationships $ds/dt=-v$ and $ds/dt=ds/dy\times dy/dt$, one can find out that

$$\{1+ [\psi'(y)]^2\}^{1/2}\, dy/dt = -[2g(Y-y)]^{1/2}. \tag{1.25}$$

Finally, the integration of (1.25) from Y to 0 with respect to y and from 0 to $f(Y)$ with respect to t leads to the following integral relation:

$$\int_0^Y \frac{\phi(y)}{(Y-y)^{1/2}}\,dy = f(Y), \quad 0\leq Y\leq b, \tag{1.26}$$

where

$$\phi(y) = \{[1 + (\psi'(y))^2]/2g\}^{1/2}. \qquad (1.27)$$

The formulas (1.26)-(1.27) solve the direct problem of finding out the sought-for descent time $f(y)$ under the given $\psi(y)$, $0 \le y \le b$. For the inverse problem (finding the unknown curve shape $\psi(y)$ under the given descent time $f(y)$), the formula (1.26) represents the Volterra integral equation of the first kind with respect to $\phi(y)$, $0 \le y \le b$, which is called *Abel's equation.*

Since the original Abel's work of 1823, many physical problems have been discovered that lead to the Abel integral equation (1.26) and its modifications. For example, the problem of rolling a ball up a hill of unknown shape is described by the same equation (1.26) (Logan, 1987). The common generalization of the Abel integral equation (1.26) is the *generalized Abel equation*:

$$\int_0^x (x-t)^{-\alpha}\phi(t)dt = f(x), \quad 0 \le \alpha \le 1, \quad 0 \le x \le 1, \qquad (1.28)$$

with the solution:

$$\phi(x) = -\pi^{-1}\sin\alpha\pi \frac{d}{dt}\int_0^x (x-t)^{-\alpha} f(t)dt. \qquad (1.29)$$

2.2 Sliding Chain

A perfectly flexible chain of length a is considered, which is sliding down horizontal table along a smooth curve under gravity, without friction, as shown in Figure 1.6.

Let us introduce three key functions:

- the linear mass density $f(s)$ of the chain,
- the equation $y=\phi(s)$ of the curve OC (where s is the length of the arc),
- the length $\sigma(t)$ of the curve OA between the origin O and the end A of the chain. The function $\sigma(t)$ completely defines the position of the chain.

As shown in (Corduneanu, 1991) using the energy balance, the equation of motion of the chain has the following integral form:

$$\int_0^{\sigma} \phi'(\sigma-\lambda) f(\lambda) d\lambda = M \frac{d^2\sigma}{dt^2}, \quad 0 \le \sigma \le a. \tag{1.30}$$

where $M = \int_0^a f(\lambda) d\lambda$ is the total mass of the chain.

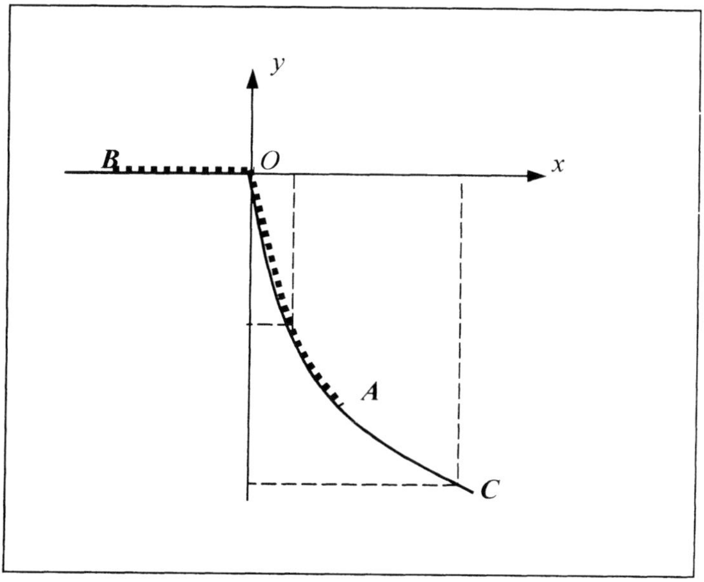

Figure 1.6. Dynamics of the flexible chain *AB* sliding of the table along the curve *OC* is described by the integral equation (1.30)

The formula (1.30) gives an integral-differential equation with respect to $\sigma(t)$ for the *direct problem* of finding $\sigma(t)$ under the given functions $f(s)$ and $\phi(s)$, $0 \le s \le a$.

For the *inverse problems* of finding the unknown chain mass density $f(y)$ or the unknown curve shape $\psi(y)$ under the given law of the motion $\sigma(t)$, the formula (1.26) delivers *the Volterra integral equation of the first kind* with respect $f(s)$ or $\phi'(s)$, $0 \le s \le a$.

When the whole chain slid off the table, i.e., $\sigma > a$, the constant upper limit of integration in the equation of motion (1.30) becomes constant and equals a (instead of the variable limit σ). Correspondingly, the above-mentioned inverse problems are described by Fredholm integral equations of the first kind, which may not possess a solution or have multiple solutions.

2.3 Models of Computerized Tomography

Tomography is an applied engineering discipline whose importance has been significantly increasing during the last two decades with the development of computers and computer simulation. The main application of tomography methods is medicine but such methods are also widely used in construction and civil engineering, non-destructive material analysis, and so on. Mathematical problems of tomography clearly demonstrate the practical importance of applied mathematics.

The basic problem of tomography lies in the reconstruction of a three-dimensional image on the basis of its two-dimensional projected distributions (scans, penetrations, profiles) taken at various angles. A mathematical tool for description of the above problem (Radon transform) was created by Johann Radon in 1917, a long time ago before the computerized tomography was born.

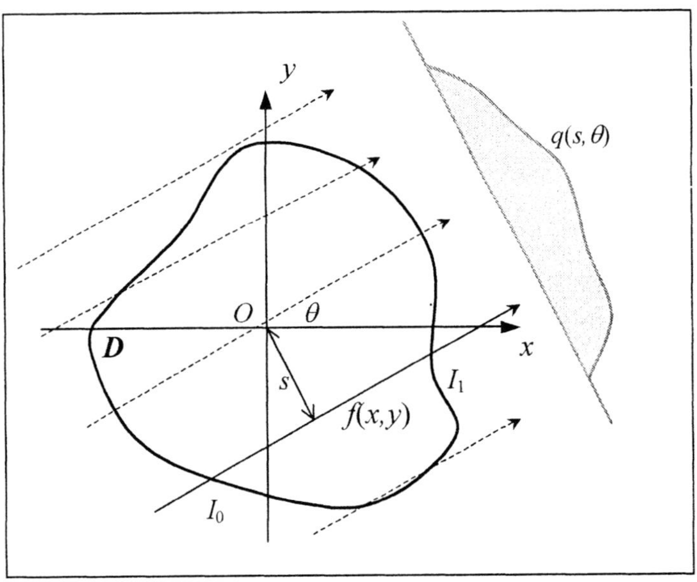

Figure 1.7. A scheme of tomography process that produces a set of projections $q(s, \theta)$ of at the angle θ of the body with the density $f(x,y)$; the grayed part illustrates the recording device for the projection distributions $q(s, \theta)$.

2.3.1 Radon Transform

A simple scheme of the tomography process is shown in the Figure 1.7. For the sake of simplicity, let us restrict ourselves with two-dimensional case. The beam source device sends a large number of proton beams (X-rays or other non-destructive penetrating radiation) across a section of the body under investigation (for example, a human brain). The sought-for characteristic of the body is its *density* (also called the *adsorption coefficient* or *image intensity*) $f(x,y)$ at the point (x,y). Let $I(x,y)$ be the (measured) beam intensity, then the relative beam transmission along a straight line L is equal to (Saaty, 1981):

$$\ln(I_1 / I_0) = -\int_L f(x,y)ds. \qquad (1.31)$$

where I_0 and I_1 are the beam intensities at the entry and exit points of the path L respectively. So, one can measure the intensities I_0 and I_1 in front of and behind the body and find out the value of the function $R(L)=\ln(I_1/I_0)$. If the function $R(L)$ is known for *all* possible lines L crossing the body, then the Radon transform technique can be used for finding the density $f(x,y)$.

The straight line L is characterized by two scalar parameters. If one selects the line slope p and its intercept τ as parameters, then the *Radon transform* of the image intensity $f(x,y)$ is defined as

$$R(p,\tau) = -\int_{-\infty}^{\infty} f(x,\tau + px)dx =$$

$$= \int_{-\infty}^{\infty}\int_{-\infty}^{\infty} f(x,y)\delta[y-(\tau + px)]dydx. \qquad (1.32)$$

The Dirac delta function in (1.32) picks rectilinear paths that are parallel to the projection line $y=px+\tau$.

The formula (1.32) represents the original classic form of Radon transform. However, the following form of Radon transform is more convenient and often used in applications:

$$q(s,\theta) = \int_{-\infty}^{\infty}\int_{-\infty}^{\infty} f(x,y)\delta[s - x\cos\theta - y\sin\theta]dydx. \qquad (1.33)$$

where the parameters are the distance s of the projection line from the origin and its angle θ off the x-axis (as shown in Figure 1.7).

Despite the presence of infinite integrals in (1.32) and (1.33), the formulas are convenient for practical use. One can easily calculate these integrals because all real-life objects have a finite size, so the unknown density $f(x,y)$ vanishes outside some finite region D. So, applied problems are reduced to reconstruction of the body density $f(x,y)$, $(x,y)\in D$, on the basis of the measured projection functions (line intergals) $q(s,\theta)$, given in the finite area $0\le s\le \max\rho(0,D)$, $0\le\theta\le 2\pi$.

The formulas (1.32) and (1.33) may be considered as linear integral equations with respect to the sought-for function $f(x,y)$), $(x,y)\in D$. Such integral equations belong to a special class called *integral transforms* (the most famous is Fourier transform). The theory of integral transforms has been well developed and existing problems are related to numerical stability and efficiency.

Thus, the solution of the "classical" Radon transform (1.32) is delivered by the following *inverse Radon transform* (Anger and Portenier, 1992; Deans, 1983):

$$f(x,y) = \frac{1}{2\pi}\int_{-\infty}^{\infty}\frac{d}{dy}H[R(p,y-px)]dp, \qquad (1.34)$$

where $H[R]$ stands for *Hilbert integral transform*:

$$H[f(X)] = h(y) = \frac{1}{\pi}\int_{-\infty}^{\infty}\frac{f(x)}{y-x}dx \qquad (1.35)$$

Despite its analytical simplicity, the inverse Radon transform (1.34) leads to computational problems connected with numerical differentiation in the formula (1.34) and singularities in (1.35).

There are several *inversion formulas* for Radon transform and the most used one is based on Fourier transform. Namely, let us compute the one-dimensional Fourier transform of the Radon transform $q(s,\theta)$ given by (1.33):

$$Q(\theta,\omega) = \int_{-\infty}^{\infty} e^{-j\omega s} q(s,\theta)ds =$$

$$= \int_{-\infty}^{\infty}\int_{-\infty}^{\infty}\int_{-\infty}^{\infty} f(x,y)e^{-j\omega s}\delta[s - x\cos\theta - y\sin\theta]dxdyds =$$

$$= \int_{-\infty}^{\infty}\int_{-\infty}^{\infty} f(x,y)e^{-j\omega(x\cos\theta + y\sin\theta)}dxdy \tag{1.36}$$

It is easy to see that the final formula (1.36) is the standard *two-dimensional Fourier transform*

$$F(u,v) = \int_{-\infty}^{\infty}\int_{-\infty}^{\infty} f(x,y)e^{-j(ux+vy)}dxdy,$$

which is measured at the position $(u,v)=(\omega\cos\theta, \omega\sin\theta)$ of Fourier transform space. Hence, one can apply the *inverse two-dimensional Fourier transform* to resolve the transform (1.36):

$$f(x,y) = \frac{1}{4\pi^2}\int_{-\infty}^{\infty}\int_{-\infty}^{\infty} F(u,v)e^{-j(ux+vy)}dxdy.$$

Namely, let us change the rectangular coordinates (x,y) to the polar coordinates (θ,ω) and replace $F(\theta,\omega)$ with $G(\theta,\omega)$. It leads to the following *inversion formula*:

$$f(x,y) = \frac{1}{2\pi^2}\int_{0}^{\pi}\int_{-\infty}^{\infty} Q(\theta,\omega)e^{-j\omega(x\cos\theta + y\sin\theta)} \,|\,\omega\,|\, d\omega d\theta, \tag{1.37}$$

where $|\,\omega\,|$ is the determinant of Jacobian of the change of variables from rectangular to polar coordinates and $Q(\theta,\omega)$ is the one-dimensional Fourier transform of the Radon transform $q(s,\theta)$.

The important practical advantage of the inversion formula (1.37) is the use of Fourier transforms. Fourier transforms can be very rapidly computed by special numeric techniques (*discrete Fourier transform, fast Fourier transform*, *z*-transform, and so on). Additionally, Fourier transform methods are very stable with respect to the *discretization errors* that arise when discrete approximations of the given functions are used (Gindikin, 1994).

In conclusion, the authors note that, because of its nature, Radon transform accentuates linear features of an image, which may cause additional errors for certain applications, see (Kunyansky, 1992; Copeland, Ravichandran, and Trivedi, 1994) for details.

2.3.2 Inverse Scattering Problems

In the tomography scheme described above, two signals are interacting: the incoming signal and transmitted signal. In other practical applications (geophysics, solar and radar engineering, astronomy and others), three different signals are essential (Ramm, 1980):

a) an incoming *incident wave* acting upon an object in some medium,
b) *a transmitted wave* that is passed through the object,
c) *a reflected wave* that is reflected by the object.

Then, the *direct scattering problem* lies in the determination of the reflected and transmitted waves when properties of the incident wave and the object itself are known. The *inverse scattering problems* are often more important and consist of the determination of the properties of the object under investigation when the incident, reflected and transmitted waves are known. For example (Logan, 1987), in geological explorations, occasional earthquakes and intentional explosions on the earth surface produce waves that, being reflected from underground layers, indicate changes in the earth density (oil reserves or other underground phenomena). In radar and solar devices, a known incident wave and an observed reflected wave are used to discover the presence of an aircraft or a submarine.

The most impressive applications of inverse scattering methods are connected with the noninvasive evaluation of internal structure in medical imaging and environmental systems. The corresponding inverse scattering problems calculate the impact of nonionizing electromagnetic radiation on biological objects. Such methods are used to produce estimates of the adsorbed energy that a biological object would receive when located in proximity to an electromagnetic radiation source.

Mathematically, the inverse scattering problems involve ordinary or partial differential equations, multi-dimensional integral transforms and integral equations (Ramm and Katsevich, 1996; Anger and Portenier, 1992; Deans, 1983; Gindikin, 1994; and others).

Chapter 2

Models Of Continuum Mechanical Systems

The mathematical models of continuum mechanics (solid, fluid and gas dynamics) represent one of the most important techniques of applied mathematics (Courant and Hilbert, 1953; Whitham, 1974; Smoller, 1983; Lin and Segel, 1974; Brekhovskikh and Goncharov, 1985; Stanislav, 1982; Landau and Lifschitz, 1986; Colton and Kress, 1992; Achenbach, 1984). These models are the source of many fundamental partial differential equations such as the heat equation, the Laplace and Poisson equations, the diffusion equation, the wave equation, and others. Modelling of continuum mechanical systems provides a source of important techniques and examples of applied mathematics more numerous than any other area of engineering. Such phenomena as elasticity, viscosity, diffusion, dispersion, wave propagation, refraction, scattering and reflection naturally occur there.

The governing equations of continuum mechanics may be derived in different ways. In this chapter, the authors consider a universal way for developing models of continuum mechanics, which is based on physical *conservation laws* such as balances of mass, momentum, and energy. These conservation laws will be written in their integral and differential forms. The integral form of conservation laws is more fundamental and general, and it is especially useful in cases when the differential form does not apply (see Section 2.5).

The authors provide the basic conservation laws and briefly analyze the resulting equations of motion for one-dimensional continuum media (Sections 1 and 2) and three-dimensional media (Section 3). More detailed investigation may be found in (Lin & Segel, 1974); see also (Logan, 1987; Svobodny, 1998). The basic example is the model of the elastic motion of a thin bar, but the fluid and gas dynamics are also discussed.

Deriving the equations of motion does not yet provide a workable model of a specific applied problem. The researcher needs to define the main goals of modelling, to decide which of the model's relations are relevant and which are not, and to determine possible constraints and control influences, initial and boundary conditions, etc.

To illustrate the specifics of applied mathematical modelling, in Section 4 the authors consider more complex models of water transport and contamination used for the control and conservation of water resources and flood prediction (in particular, for forecasting radioactive contamination of Ukrainian rivers after the Chernobyl nuclear disaster). These applied models are based on the three-dimensional continuum motion equations derived in Section 3 and take the following hydrophysical processes into account: wind and water currents, water stratification, drift of suspended particles, drift of involved particles, the pollution transfer in solute and suspense, and the pollution accumulation in bottom sediments. A stationary flow problem is investigated for a two-dimensional version of the models. Such problems appear in the prediction of spring floods.

1. CONSERVATION LAWS IN ONE-DIMENSIONAL MEDIUM

Conservation laws are fundamental physical principles. They apply to a selected quantity whose distribution in space and time characterizes the phenomenon under study (Lin and Segel, 1974). Examples of such a quantity are the mass density of fluids or solids, car density on the road, the concentration of some chemical in a fluid, and so on. The quantity in some region of space will be conserved if not moved or acted upon by some agent (chemical, molecular, electromagnetic, etc.).

Therefore, the general mechanism that governs the behaviour of a continuum physical medium is the following ***general law of balance***:

The total rate of change of a quantity in a space domain =

= the rate of quantity flux through the boundary +

+ the rate of quantity creation within the domain

In this section the authors consider the conservation laws in a one-dimensional physical continuum (solid or fluid). The three-dimensional case is discussed in Section 3.

1.1 Eulerian and Lagrangian Coordinates

There are two basic coordinate systems used for description of the motion of a continuum medium:

a) *Eulerian* (or *spatial*) *coordinate* x that picks out a particular location in space;
b) *Lagrangian* (or *material*) *coordinate* η that picks out a specific particle of the moving continuum (cross section of a solid bar).

As time evolves, the particle that was initially at the position η is now at

$$x = x(\eta, t). \tag{2.1}$$

The inverse relation is also valid: $\eta = \eta(x, t)$.

It is important to distinguish these two coordinates. We will use lowercase letters for functions of the spatial variable (*Eulerian quantities*) and uppercase letters for functions of the material variable (*Lagrangian quantities*). Thus, if $F(\eta, t)$ is the Lagrangian quantity of some characteristic of the medium, then the Eulerian quantity of this characteristic is described as $f(x, t) = F(\eta(x, t), t)$ and, inverse, $F(\eta, t) = f(x(\eta, t), t)$.

The difference between the current position x and the initial position η of a particle is known as the *displacement*:

- $U(\eta, t) = x(\eta, t) - \eta$ – in material coordinates,
- $u(x, t) = U(\eta(x, t), t) = x - \eta(x, t)$ – in spatial coordinates.

As in classical mechanics, the *velocity* of a particle is defined as the time rate of its position change:

- $V(\eta, t) = \partial x(\eta, t)/\partial t = \partial U(\eta, t)/\partial t$ – in material coordinates,
- $v(x, t) = V(\eta(x, t), t)$ – in spatial coordinates.

The Jacobian of the one-dimensional transformation (2.1) is

$$J(\eta, t) = \partial x(\eta, t)/\partial \eta \tag{2.2}$$

and its derivative is defined by the *Euler expansion formula*:

$$\partial J(\eta,t)/\partial t = J(\eta,t)\ \partial v(x,t)\ /\partial x \tag{2.3}$$

Then, using the chain rule, one can define the so-called *material* (or *convective* or *full*) *derivative* Df/Dt of the function $f(x, t)$ as

$$Df(x,t)/Dt \ = \partial f(x,\ t)/\partial t + v(x,\ t)\partial f(x,\ t)/\partial x. \tag{2.4}$$

It describes the rate of the change of the function f following a particle η that now is located at x (as seen by an observer who is riding with the particle).

1.2 Mass Conservation

Let us denote the amount of a conserved quantity per unit volume at the position x as $\rho(x,t)$ in the spatial coordinates (the *Eulerian mass density*) and use the notation

$$R(\eta,\ t) = \rho(x(\eta,t),\ t)$$

for the quantity per unit volume in material coordinates (the *Lagrangean mass density*).

We consider an arbitrary material portion (a box or cylinder) that moves with the underlying material. If the box was $A\times[a_0, b_0]$ at t=0, then in accordance with (2.1) this portion of material has moved to the region between $x=a(t)=x(a_0,t)$ and $x=b(t)=x(b_0,t)$.

We assume that the mass of the box does not change as it moves in time:

$$\frac{d}{dt}\int_{a(t)}^{b(t)}\rho(x,t)dx = 0. \tag{2.5}$$

Using the change of variables (2.1) and the formula (2.4), one can obtain

$$\frac{d}{dt}\int_{a_0}^{b_0}R(\eta,t)J(\eta,t)d\eta = 0.$$

Now, taking the derivative of the integrand and using the Euler expansion formula (2.3) lead to:

$$\int_{a_0}^{b_0}[\partial R(\eta,t)/\partial t J(\eta,t) + R(\eta,t)\partial J(\eta,t)/\partial t]d\eta =$$

$$= \int_{a_0}^{b_0}[\partial R(\eta,t)/\partial t + R(\eta,t)\partial v(\eta,t)/\partial x]J(\eta,t)d\eta = 0.$$

Returning to the spatial coordinates in the last formula, we obtain the following *integral law of mass conservation*:

$$\int_{a(t)}^{b(t)}[D\rho(x,t)/Dt + \rho(x,t)\partial v(x,t)/\partial t]dx = 0. \tag{2.6}$$

Assuming that the intergrand in (2.6) is continuous, one can apply the Dubois-Reymond lemma (see Appendix) to (2.6) and obtain:

$$D\rho/Dt + \rho\,\partial v/\partial x = 0 \tag{2.7}$$

or

$$\partial\rho/\partial t + v\,\partial\rho/\partial x + \rho\,\partial v/\partial x = 0 \tag{2.8}$$

The equation (2.8) is called the *continuity equation*.

1.3 Momentum Conservation

The momentum balance is derived from the **Newton's second and third laws of mechanics**. Applied to a one dimensional continuum medium, the **Newton's second law** states that the rate of change of linear momentum of any portion of material is equal to the sum of external forces acting upon it (see also Section 1.1 of previous Chapter).

The linear momentum of the portion of a material with density $\rho(x,t)$ in the region (the box) $A\times[a(t),b(t)]$ between $x=a(t)$ and $x=b(t)$ is defined by

$$\mathbf{i}A\int_{a(t)}^{b(t)}\rho(x,t)v(x,t)dx\,,$$

where

i is the unit vector of the x-axis.

$v\ (x,\ t)$ is the velocity of medium in spatial coordinates (see Section 1.1).

There are two types of external forces that act on the material region:

a) *body forces* that act on the material particle and are proportional to its mass (like gravity). We will denote the body force vector per unit mass as $\mathbf{f}(x,t) = \mathbf{i}\, f(x,t)$.

b) *surface forces* that act across surfaces (like pressure). Let us introduce the *stress vectors* $\boldsymbol{\sigma}(x,t,\mathbf{i})$ and $\boldsymbol{\sigma}(x,t,\text{-}\mathbf{i})$ for describing the surface forces. The surface force $\boldsymbol{\sigma}(x,t,\mathbf{i})$ per unit area acts across a section located at x at time t. This force is exerted *by* the material on the positive side *on* the material on the negative side. The unit surface force $\boldsymbol{\sigma}(x,t,\text{-}\mathbf{i})$ across a section x is exerted *by* the material on the negative side *on* the material on the positive side.

Then the vector balance of linear momentum for a material region of the volume $A\times[a(t),b(t)]$ is

$$\mathbf{i}\frac{d}{dt}A\int_{a(t)}^{b(t)}\rho(x,t)v(x,t)dx =$$

$$= A\int_{a(t)}^{b(t)}\rho(x,t)\mathbf{f}(x,t)dx + A\boldsymbol{\sigma}(b(t),t,\mathbf{i}) + A\boldsymbol{\sigma}(a(t),t,-\mathbf{i}) \qquad (2.9)$$

A continuum version of the **Newton's third law of mechanics** states that the action is equal to the reaction or

$$\boldsymbol{\sigma}(x,t,\ -\mathbf{i}) = -\ \boldsymbol{\sigma}(x,t,\mathbf{i}).$$

This relation allows us to introduce the scalar *stress component* $\sigma(x,t)$ as

$$\sigma(x,t,\mathbf{i}) = \mathbf{i}\sigma(x,t).$$

Then the momentum balance equation may be written in a scalar form as:

$$\frac{d}{dt}\int_{a(t)}^{b(t)}\rho(x,t)v(x,t)dx =$$

$$= \int_{a(t)}^{b(t)}\rho(x,t)f(x,t)dx + \int_{a(t)}^{b(t)}\partial\sigma(x,t)/\partial x dx \qquad (2.10)$$

Using the same transformations as for the mass conservation law (changing to Lagrangian coordinates and back, using Jacobian and the formula (2.3)), we obtain the following *integral form of the momentum conservation law* :

$$\int_{a(t)}^{b(t)}[\rho\partial v/\partial t+\rho v\partial v/\partial x-\partial\sigma(x,t)/\partial x-\rho f]dx=0 \qquad (2.11)$$

Assuming the continuous intergrand and using the Dubois-Reymond lemma from Appendix (Chapter 7), we obtain the *differential form of the momentum conservation law* :

$$\partial v/\partial t+\ v\,\partial v/\partial x-\rho^{-1}\partial\sigma/\partial x=f. \qquad (2.12)$$

The conservation equations (2.8) and (2.12) are valid for any continuum medium, but they do not provide a complete mathematical model of the corresponding physical process. So we need one more equation to complete the model of the process. The two equations (2.8),(2.12) connect three independent sought-for variables: density ρ, velocity v (or displacement u, $v=Du/Dt$) and stress σ. The additional equation is usually based on the so-called *constitutive equations* or *phenomenological equations* that specify the particular nature and properties of a specific medium (in contrast with universal conservation laws, which describe the behaviour of all materials). The constitutive equations may reflect experimental observations.

Thus, for a one-dimensional motion of a solid bar (slender) considered below in Section 2, a constitutive relation defines the connection of the distortion (strain) with the amount of stress that produces this distortion. In the simplest case, this is Hook's law. This constitutive relation reflects the particular properties of material.

In general case, the constitutive relations use thermodynamic variables such as pressure, energy, temperature, and others. One of the most important constitutive relations is a state equation. The *state equation* describes the thermodynamic state of the material being considered; in particular, it describes whether the state is solid or fluid.

1.4 Energy Conservation and Thermodynamics

In accordance with the **Newton's second law of mechanics**, the equation of the motion of a particle is $d(m\mathbf{v})/dt=\mathbf{f}$, where **f** is the force acting on the particle and m is its mass. We define the work done by the force acting on a particle going from point a to b as a scalar value

$$W_{ab} = \int_a^b \mathbf{f}d\mathbf{s} = \int_a^b m\frac{d\mathbf{v}}{dt}\frac{d\mathbf{s}}{dt}dt = \int_a^b m\frac{d\mathbf{v}}{dt}\mathbf{v}dt =$$

$$= \frac{1}{2}m\int_a^b \frac{d(v^2)}{dt}dt = \frac{1}{2}mv^2(b) - \frac{1}{2}mv^2(a).$$

The quantity $T = \frac{1}{2} mv^2$ is called the *kinetic energy* of a particle. The change in kinetic energy is equal to the work done.

The force **f** is called *conservative* if its work done around any closed contour is zero. For a conservative force, one can define the *potential energy* of a particle U as a scalar-valued function of the (vector) coordinates. Setting the energy at some point equal zero, the energy at any other point is equal to the negative value of the work done by the force in moving from the zero point to the new point.

For a one-dimensional continuum of the length L, we can define the kinetic energy and the potential (or internal) energy as (Vierck, 1967; Lin and Segel, 1974):

$$T = \frac{1}{2}\int_0^L \rho\left(\frac{du}{dt}\right)^2 dx, \quad U = \frac{1}{2}\int_0^L E\left(\frac{du}{dx}\right)^2 dx + C,$$

where the potential energy T is determined by the work done by the elastic forces in returning to a neutral configuration of the continuum system, ρ is a "linear density" (per a length unit) and E is a "linear tension".

If all forces acting on the system are conservative, than the total energy, $T+U$, is conserved. If a system is subjected to frictional and other dissipative forces that generate heat, then the system is not mechanically conservative (its total mechanical energy decreases in time). Then we need to involve thermodynamic concepts of the process into analysis.

The **first law of thermodynamics** states that the total energy is conserved if the heat is taken into account (Logan, 1987) : the change ΔE of the internal energy E of a mechanical system equals to $\Delta E = Q - W$, where

Q is a heat added to the system and W is the work done by the system in passing from an initial state to a final one.

This law is valid for an equilibrium state only (when velocity and kinetic energies are zero). The effect of motion is that the total energy equals to the sum of the internal energy and the energy of motion (the kinetic energy). The kinetic energy is due to the gross macroscopic motion of the medium while the internal energy reflects the microscopic motion of molecules.

In a continuum medium (solid or fluid) we use the following distributed thermodynamic state variables (per unit volume): specific pressure p, specific absolute temperature θ, specific internal energy e, specific entropy s.

Following (Lin and Segel, 1974), we can postulate the balance of energy law in integral form, consistent with our previous balances. The balance states that the change of the total energy is equal to the total work that all the forces (body forces f and surface forces σ) do on the region plus the heat flux ϕ into the region. Providing our standard reasoning as for mass and momentum conservation laws, we obtain the *balance of energy* in the integral form:

$$\int_{a(t)}^{b(t)}\left[\frac{1}{2}\rho\frac{Dv^2}{Dt}+\rho\frac{De}{Dt}-\rho fv-\frac{\partial\sigma(x,t)}{\partial x}+\frac{\partial\phi}{\partial x}\right]dx=0. \qquad (2.13)$$

and in the differential form:

$$\frac{1}{2}\rho\frac{Dv^2}{Dt}+\rho\frac{De}{Dt}=\rho fv+\frac{\partial\sigma(x,t)}{\partial x}-\frac{\partial\phi}{\partial x} \qquad (2.14)$$

Another local energy conservation relation in the form of partial differential equation comes from a combination of the first and second laws of thermodynamics. It is the so-called *Gibbs relation* :

$$\frac{De}{Dt}=\theta\frac{Ds}{Dt}-p\frac{Ds}{Dt}(1/\rho), \qquad (2.15)$$

where s is the entropy.

Combining two local energy conservation principles (2.14) and (2.15), one can obtain a relation between the entropy change and the heat flux:

$$\theta \frac{Ds}{Dt} = -\frac{1}{\rho}\frac{\partial \phi}{\partial x}. \tag{2.16}$$

Then one can exclude the entropy s from the energy equation (2.15):

$$\frac{De}{Dt} + p\frac{D(1/\rho)}{Dt} = -\frac{1}{\rho}\frac{\partial \phi}{\partial x}. \tag{2.17}$$

Thermodynamic relations similar to the energy equation (2.17) are particularly important in fluid and gas dynamics. They allow deriving various models of motion under different assumptions.

A. ADIABATIC PROCESSES.

Let us assume that the entropy is not changed during the process: $Ds/Dt=0$. By virtue of (2.16), it means that we ignore the heat generation due to viscosity (the heat flux is zero). Then we have:

$$\frac{De}{Dt} + p\frac{D(1/\rho)}{Dt} = 0, \tag{2.18}$$

and the entropy s may be excluded from the set of model equations.

B. ADIABATIC PROCESSES IN IDEAL GASES.

Under normal conditions most gases obey the so-called *ideal gas law*:

$$p = R\rho\theta, \qquad e = c_v\theta + \text{const}, \tag{2.19}$$

where the parameter c_v is the specific heat at constant volume and R is the gas constant. If we consider adiabatic flow of the ideal gas, then the last two equations together with the energy equation (2.17) and the differential balance laws (2.8) and (2.12) form a set of five equations for the five unknowns ρ, v, p, θ, and e, so we have a complete mathematical model of the process. However, these equations are nonlinear and can not be resolved in a general case. So, the further assumptions are needed.

C. PROPAGATION OF SOUND.

The so-called *acoustic approximation* is a specific case of the adiabatic flow of ideal gas and describes the process of sound propagation. It is based on a linearization of the previous set of equations under the assumption that the deviations of the unknown functions ρ, v, p, from a constant equilibrium state $v_0=0$, ρ_0, and p_0 are small. Then the linearized model of the process is described by one PDE (wave equation):

$$\delta_{tt} - c_0^2\, \delta_{xx} = 0, \tag{2.20}$$

with respect to the relative changes $\delta=(\rho-\rho_0)/\rho$ of the density ρ, where the parameter c_0 is the speed of acoustic waves propagation.

One can find more detailed and complex acoustic models in (Filippi et al, 1999) and many other textbooks on acoustics. We will consider one applied engineering acoustic problem and solve it using the boundary integral equation method in Section 2 of Chapter 4.

For a general nonadiabatic flow involving heat conduction, we need an additional constitutive relationship between the heat flux and temperature. In particular, using the empirical Fourier's heat law leads to diffusion processes. This part is described in more detail in Section 3 for a three-dimensional continuum.

2. MODELS OF ONE-DIMENSIONAL CONTINUUM DYNAMICS

In Section 1 we have derived *field equations* that describe general physical laws of mass and momentum conservation. These equations are valid for any continuum medium. On the other hand they do not constitute a complete mathematical model. To proceed further in the modelling, we need to choose a more specific case. In this section we consider application of the one-dimensional conservation laws in solid mechanics. Our basic physical model is a thin cylindrical bar (slender). We will investigate the *longitudinal dynamics* of such a bar, which corresponds to small longitudinal distortions (compression and elongations) of the bar under the influence of external forces. Dissemination of the distortions through the medium leads to stress waves in the bar. Such waves may be considered as a generalization of mechanical vibrations of Chapter 1 for the continuum mechanics case. A mathematical description of the distortion, stress and strain gives an additional *constitutive relation* that will complete the mathematical model of the process.

2.1 Nonlinear Model of Solid Bar Dynamics

Following (Lin and Siegel, 1974), a reasonable measure of the one-dimensional distortion is the difference between a new length and original length of a small bar portion divided by the original length ΔL of the portion. For small ΔL , this measure is well approximated by the *strain*

$$Z(A,t) = \partial U(A,t)/\partial A, \tag{2.21}$$

where $U(A,t)$ is the displacement of a bar section A at time t (in material coordinates, see also Section 1.1). In a one-dimensional medium, the strain may be considered as a change in infinitesimal length per unit original infinitesimal length.

The field equations (2.8) and (2.12) bring together three sought-for variables: density ρ, velocity v (or displacement u, $v=Du/Dt$), and stress component σ. So, the missing constitutive relation should connect the distortion (strain) with the amount of stress that produces this distortion. The simplest linear version of such relation is called the **Hook's law**:

$$\Sigma(A,t) = E(A)Z(A,t), \tag{2.22}$$

where $\Sigma(A,t)$ denotes the stress in material coordinates and the proportionality factor $E(A)$ is called *Young's modulus* or the *stiffness* and is determined by particular properties of the bar material.

In spatial (Eulerian) coordinates the Hook's law becomes

$$\sigma(x,t) = E(x)\zeta(x,t). \tag{2.23}$$

where $\zeta(x,t)$ denotes the Eulerian presentation of the strain (2.21). In virtue of the relation

$$\left.\frac{\partial U}{\partial A}\right|_{A=A(x,t)} = \frac{\partial u(x,t)/\partial x}{1-\partial u(x,t)/\partial x},$$

between the displacements U and u in material and spatial coordinates (see Section 1.1), we obtain the following constitutive equation:

$$\sigma(x,t) = E(x)\frac{\partial u(x,t)/\partial x}{1-\partial u(x,t)/\partial x}. \tag{2.24}$$

The Hook's law reflects our intuitive feeling that the larger force applies, the more bar elongates. However, this constitutive relation is valid for small stress values only and is not valid for some specific materials. Possible generalizations of the Hook's law lead to more complicated mathematical models (viscoelasticity) considered in Section 2.5.

Thus, we have three nonlinear equations (2.8), (2.12) and (2.24) with respect to sought-for displacement $u(x,t)$, stress $\sigma(x,t)$, and density $\rho(x,t)$. Using (2.24) and the relation $v=Du/Dt$, we can eliminate the stress and velocity from the momentum conservation equation (2.12) and write it as

$$\rho\frac{\partial^2 u}{\partial t^2} = \rho f + \frac{\partial}{\partial x}\frac{E\partial u/\partial x}{1-\partial u/\partial x} \tag{2.25}$$

with respect to $u(x,t)$ and $\rho(x,t)$. These nonlinear equations may be investigated by various numeric and analytic methods.

2.2 Linearized Model of Solid Bar Dynamics

A common analytic technique consists of linearization of the nonlinear equations (2.8), (2.12) and (2.24) of a solid bar dynamics for the case of small displacements $|u|<<1$. During the linearization process, the system (2.8), (2.12) and (2.24) is reduced to a single equation in one variable. Namely, the linearized momentum equation

$$\rho(x)\frac{\partial^2 u}{\partial t^2} = \rho(x)f(x) + \frac{\partial}{\partial x}\left[E(x)\frac{\partial u}{\partial x}\right] \tag{2.26}$$

has the single unknown $u(x,t)$. Here $\rho(x)$ means the (given) initial density and $E(x)$ is the initial stiffness.

Once $u(x,t)$ is determined from (2.26), then (in the first approximation) the sought-for stress $\sigma(x,t)=E(x)\partial u(x,t)/\partial t$ and the corrected sought-for density function $\rho(x,t)$ may be computed from the following linearized mass conservation equation:

$$\frac{\partial}{\partial t}\left[\rho(x,t)-\rho(x)\right]=-\frac{\partial}{\partial x}\left[\rho(x)\frac{\partial u}{\partial t}\right] \tag{2.27}$$

The equation (2.27) is usually ignored in many applications. Sometimes it serves as a criterion for checking the smallness of the sought-for functions.

The complete linearized mathematical model of longitudinal motion in a solid bar of the final length L consists of partial differential equation (2.26) of the second order with respect to the unknown $u(x,t)$, $0 \leq x \leq L$, $t \geq 0$, with the following initial conditions at t=0:

$$u(x,\ 0)= u_0(x), \qquad \partial u(x,\ 0)/\partial t = v_0(x), \tag{2.28}$$

and appropriate boundary conditions at x=0 and x=L. Examples of the boundary conditions are (for the right end x=L) :

$$u(L,\ t)= 0, \qquad \text{(fixed end)} \tag{2.29a}$$

$$E(L)\partial u(L,\ t)/\partial x = 0, \qquad \text{(free end)} \tag{2.29b}$$

$$M\partial^2 u(L,t)/\partial t^2 + E(L)\partial u(L,t)/\partial x = 0, \text{ (free end with mass } M) \tag{2.29c}$$

and so on.

If the initial density and stiffness are constant, $\rho(x)\equiv\rho_0$ and $E(x)\equiv E_0$, and the external force $f\equiv 0$, then the partial differential equation (2.26) reduces to the *wave equation*:

$$u_{tt} - c^2 u_{xx} = 0, \tag{2.30}$$

where the parameter $c = (E_0/\rho_0)^{1/2}$ is called the *speed of sound.*

Different aspects of wave equation (2.30) were thoroughly investigated by many famous scientists during past decades. We shall restrict ourselves with immediate facts necessary for our further analysis and refer to famous books (Whitham, 1974; Smoller, 1983; Logan, 1987; etc.) for other known results.

It is well known that the general solution of the equation (2.30) has the form

$$u(x, t) = f(x - ct) + g(x + ct), \tag{2.31}$$

where $f(x - ct)$ and $g(x + ct)$ may be considered as *traveling waves* that move to the right and to the left with the fixed speed c. The form and behaviour of the travelling waves f and g are determined by the initial and boundary data (2.28) and (2.29) and vary widely for different cases. In fact, many famous methods of PDE theory were developed during solving particular cases of the problem (2.26)-(2.29).

The initial problem for the wave equation (2.30) with the initial condition (2.28), $x \in (-\infty, \infty)$, $t \geq 0$, has a so-called *D'Alembert solution*:

$$u(x,t) = \frac{1}{2}\left[f(x+ct) + f(x-ct)\right] + \frac{1}{2}\int_{x-ct}^{x+ct} G(y)dy \tag{2.32}$$

If the above-formulated problem of longitudinal bar dynamics is considered on a limited space interval $[0, L]$, then boundary conditions of the type (2.29) should be presented. The dynamics in boundary problems is more complicated as compared with the initial problem dynamics (2.32) because waves are reflected from the boundaries (the functions f and g in (2.31) are composed from the direct and reflected waves).

We will briefly analyze some generalizations of the linear models and classic techniques that appear in various applied engineering problems. In doing so, we will concentrate on real life situations where some assumptions of classic analysis (smoothness of given functions in the model equations, small amplitudes, proportionality of stress and strain, etc.) are not valid. For example, real engineering systems are usually constructed from separate pieces of different materials and shapes and are subjected to sudden loads. Therefore, investigation of possible discontinuities in the model solutions plays a special role in applied analysis.

A basic classification of discontinuities in continuum media includes the following two cases: *shock waves* and *interfaces*. Shock waves are characterized by the fact that matter crosses the discontinuity surface. In the case of an interface between two immiscible physical media, the matter does not cross the discontinuity surface (Filippi et al, 1999). The interfaces may be solid-solid (adhesive and sliding interfaces), solid-fluid, fluid-fluid, and gas-fluid. The dynamics of interfaces is often described mathematically as *moving boundary problems*.

The discontinuous waves in continuum appear because of the following major reasons:

a) Physical and geometric properties of the medium such as density, stiffness, etc., are discontinuous (instantaneously changed) at some points x including different phases of the same material.
b) A sudden load is instantaneously applied to a region of a continuum system.
c) Some nonlinear physical phenomena, i.e., dependence of wave propagation speed on the wave amplitude, naturally lead to discontinuities in many solids, fluids and gases.

Discontinuities of the first two types are investigated in the frame of the linear wave propagation theory in Section 2.3. Self-formation of the discontinuous waves because of nonlinear effects (even when internal properties and external forces are continuous) is analyzed in Section 2.4, where the basic process is a stress wave in a one-dimensional solid bar. In all cases, non-traditional techniques based on integral conservation laws are used for the analysis of discontinuous solutions.

Section 2.5 deals with the modelling of viscous-elastic materials for which the Hook's law (2.23) about the proportionality of stress and strain is not valid.

2.3 Discontinuities in Linear Models

In this section, we will use the following generalization of the linearized momentum equation (2.26) of longitudinal bar dynamics :

$$\rho(x)S(x)\frac{\partial^2 u}{\partial t^2} = \rho(x)S(x)f(x) + \frac{\partial}{\partial x}\left[S(x)E(x)\frac{\partial u}{\partial x}\right], \tag{2.33}$$

that takes the variable cross-section area $S(x)$ of the bar into account. The derivation of the equation (2.33) is similar to the case of (2.26).

Classic analysis primarily deals with ideal situations when the given functions $\rho(x)$, $S(x)$, $E(x)$, $f(x)$ of the equation (2.33) are twice continuously differentiable and all functions in initial and boundary conditions (2.28)-(2.29) are at least continuous. Applied problems are often different. For example, let us consider a bar suddenly struck at one end. Mathematically it means that the function $f(x)$ of body force in (2.33) or one of the initial/boundary functions in (2.28)-(2.29) is discontinuous at time t. Another case arises when some properties of the bar (density, stiffness, etc.) are suddenly changed, then some of the functions $\rho(x)$, $S(x)$, $E(x)$, $f(x)$ in (2.33) are discontinuous in x. In both cases, a discontinuous solution may appear in the model. Classical tools are not functional for investigating such situations. We will consider some techniques for analysis of discontinuous solutions, based on integral conservation laws.

2.3.1 Analysis of Discontinuity Propagation

Let us assume that, because of some external impact, the stress $\sigma(x,t)=E(x)\partial u(x,t)/\partial t$ and the velocity $v(x,t)=\partial u(x,t)/\partial t$ in the model (2.33) undergo some (known) finite jumps at time $t=t_0$, but the displacement $u(x,t)$ remains continuous. This is the case of the so-called "mild" discontinuity when the bar is not permanently deformed or folded. Then the moving discontinuous solution cannot satisfy the differential equation (2.33) but the underlying integral conservation relations are still valid. The question is to determine how the *discontinuity surface* is propagated through the medium. A method of investigating this question is based on using the integral form of momentum conservation equation (2.11).

Let us assume that the position of the *moving discontinuity surface* is given by the equation

$$\phi(x, t) = 0 \qquad \text{or} \qquad x = x(t) \tag{2.34}$$

Then the speed dx/dt of the discontinuity surface is related to its shape as

$$dx/dt = -\ \phi_t(x, t)/\phi_x(x, t) \tag{2.35}$$

We will denote the jump in a function $f(x,t)$ across the discontinuity surface as

$$[f] = \lim_{\xi\to x-0} f(x,t) - \lim_{\xi\to x+0} f(x,t) \tag{2.36}$$

Since the displacement $u(x,t)$ is continuous, then $[u]=0$. We also assume that the jumps in $u_x(x,t)$ and $u_t(x,t)$ remains continuous across the discontinuity surface.

Integrating *along* the discontinuity surface $\phi(x,t)=0$ between its arbitrary points (x_0, t_0) and (x_1, t_1) gives:

$$\int_{(x_0,t_0)}^{(x_1,t_1)} [u_s(x,t)]ds = 0 \tag{2.37}$$

Then, by Dubois-Reymond lemma (see Appendix), we obtain $[u_s(x,t)]=0$, which leads to the following relation between the jumps $[u_x]$ and $[u_t]$ in the first derivatives:

$$[u_x(x,t)] = [u_t(x,t)]\phi_x(x, t)/\phi_t(x, t) \tag{2.38}$$

We need the second relationship between $[u_x]$ and $[u_t]$, which may be obtained from the model of the process. Based on Section 2.1.2, the integral form of momentum conservation equation (2.11) for the model (2.33) may be written as:

$$\iint_G \rho S u_{tt} dx dt = \oint_{\partial G} ES u_x n_1 ds,$$

or, by using the Green's theorem (Courant and Gilbert, 1953), as

$$\oint_{\partial G} (ESu_x n_1 - \rho S\, u_t n_2) ds = 0. \tag{2.39}$$

where G is an arbitrary region in the (x,t)-plane and n_1 and n_2 are components of the unit exterior normal vector $\mathbf{n}=n_1\mathbf{i}+n_2\mathbf{j}$.

Applying the last formula to an arbitrary part Γ of the discontinuity surface, after transformations we obtain that

$$\oint_\Gamma (ES[u_x]n_1 - \rho S\, [u_t]n_2) ds = 0, \tag{2.40}$$

or, by Dubois-Reymond lemma (see Appendix) :

$$ES\,[u_x]\,n_1 - \rho S\,[u_t]\,n_1 = 0 \tag{2.41}$$

along $\phi(x,t)=0$. Since the normal vector **n** is proportional to $\phi_x\mathbf{i}+\phi_t\mathbf{j}$, then:

$$ES\,[u_x]\phi_x - \rho S\,[u_t]\phi_t = 0. \tag{2.42}$$

The system of two linear homogeneous equations (2.38) and (2.42) has a nontrivial solution $\{[u_x],\ [u_t]\}$ only if its determinant is zero:

$$-E\phi_x^2 + \rho\phi_t^2 = 0 \qquad \text{or} \qquad (\phi_x/\phi_t)^2 = \rho/E.$$

Now, using (2.35), we finally obtain the following formula for the speed of the moving discontinuity surface (wave front):

$$dx/dt = \pm\, c(x), \quad c(x) = [\rho(x)/E(x)]^{1/2}, \tag{2.43}$$

where the function $c(x)$ is called *the local speed of sound.* Note that the local speed of sound depends on physical properties (the density and stiffness) and is independent of the cross-section area (geometry).

Thus, small discontinuities propagate with the local speed of sound. This result generalizes the classical one (small smooth disturbances propagate with the *fixed* speed of sound in a *uniform* bar) that follows from the formula (2.31).

2.3.2 Analysis of Size of Discontinuity

The next important question is about the dynamics of the discontinuities. The *wave function* $\psi(x)$ obtained from solving the equation (2.34) of the moving discontinuity surface $\phi(x,t)=0$ with respect to t, $t=\phi(x)$, plays an important role below. The wave function satisfies the so-called *eiconal equation*

$$\psi'^2(x) = 1/c^2(x). \tag{2.44}$$

We will consider a wave front propagating to the right. Let us denote the jump of the velocity $v(x,t)=\partial u(x,t)/\partial t$ at the discontinuity $\phi(x,t)=0$ (or $t=\psi(x)$) as

$$V(x) = [v(x,t)]\,|_{t=\psi(x)} = v(x, \psi(x)).$$

Analyzing the equation of motion (2.33) on the right- and left-hand sides of the discontinuity and expressing it in terms of $\psi(x)$, one can obtain the following equation for the velocity jump $V(x)$:

$$V'(x)/V(x) = -\tfrac{1}{2}\,[E(x)S(x)\psi'(x)]'\,/[E(x)S(x)\psi'(x)]^{2} \tag{2.45}$$

or after integrating:

$$V(x)/V(x_0) = [E(x_0)S(x_0)\psi'(x_0)]^{1/2}/[E(x)S(x)\psi'(x)]^{1/2} \tag{2.46}$$

where x_0 is the location of the discontinuity surface at time $t=t_0$. Recalling that $\psi'=1/c$ and $E=c^2\rho$, we obtain the following *propagation equation for the jump in velocity*:

$$[v(x,t)]_{t=\psi(x)} = [v(x_0,t_0)]_{t_0=\psi(x_0)}\left[\frac{\rho(x_0)S(x_0)c(x_0)}{\rho(x)S(x)c(x)}\right]^{1/2}. \tag{2.47}$$

Using the same technique, we obtain a similar propagation equation for the stress σ:

$$[\sigma(x,t)]_{t=\psi(x)} = [\sigma(x_0,t_0)]_{t_0=\psi(x_0)}\left[\frac{\rho(x)c(x)S(x_0)}{\rho(x_0)c(x_0)S(x)}\right]^{1/2}. \tag{2.48}$$

Note that the dynamics of the velocity and stress jumps depends on physical properties (the density $\rho(x)$ and stiffness $E(x)$) as well as on geometry (the cross-section area). In particular, these jumps increase when the cross-section area $S(x)$ decreases. For example, if we consider a bar of a uniform material that has the shape of a wedge, then the jumps $[v(x,t)]|_{t=\psi(x)}\to\infty$ and $[\sigma(x,t)]|_{t=\psi(x)}\to\infty$ when $S(x)\to 0$. Of course, this result is valid only for small values of v and σ (to the extent of our linearized model).

2.3.3 Discontinuities Fixed in Space

Another type of discontinuous behaviour arises when the physical or geometric properties of the bar are not smooth but have jumps. More precisely, if some of the functions of density $\rho(x)$, stiffness $E(x)$, cross-section area $S(x)$ are not continuous in the model (2.33), then the integral forms (2.6) and (2.11) of the conservation laws for mass and linear momentum still hold true. However, Dubois-Reymond lemma cannot be used and all differential conservation laws (2.8) and (2.12) are not valid anymore.

Because the location of discontinuities is known at advance and fixed in spatial coordinates, this type of discontinuity (a fixed interface surface) is simpler than considered in the previous section. However, the situation is similar and we have to rely on integral forms (2.6) and (2.11) of the conservation laws as more general as compared with their differential analogies.

A very simple but useful example of such discontinuity is considered in (Lin and Segel, 1974). It consists of an infinite bar with the only jump in the density $\rho(x)$ and stiffness $E(x)$ at $x=0$ and constant material properties for $x<0$ and $x>0$ (a *solid-solid interface* at $x=0$). A simplified analysis of this case shows that the wave propagation in such a medium includes a new phenomenon: *reflection of an incoming wave* from the discontinuity surface. More precisely, an incident wave W_I moving to the right in the region $x<0$ produces a left-moving wave W_R reflected back into the region $x<0$ by the discontinuity surface $x=0$ and a wave W_T transmitted into the region $x>0$. The speeds of the waves are defined by the corresponding properties of the region:

$$c^+ = (\rho^+/E^+)^{1/2}, \quad c^- = (\rho^-/E^-)^{1/2}.$$

Assuming that all the waves W_I, W_R, W_T are sinusoidal, the following relations between their amplitudes A_I, A_R, A_T hold true:

$$A_R = A_I\,(1-R)/(1+R), \qquad A_T = 2A_I\,/\,(1+R),$$

where the parameter $R = [(\rho^+/E^+)/(\rho^-/E^-)]^{1/2}$ is determined by the density $\rho(x)$ and stiffness $E(x)$. It is easy to see that there is no reflection wave in the case $R=1$ (uniform bar). In the case $R>>1$ (a light flexible bar connected to a heavy stiff bar), there is a standing wave in the light bar and no motion in the heavy bar. In the case $R<<1$ (a heavy stiff bar connected to a light flexible bar), the transmitted wave is of a maximum amplitude.

2.4 Discontinuities in Nonlinear Models

As follows form Section 2.3, the propagation of discontinuities can be satisfactory described in the linear theory of wave propagation. Discontinuous waves (shock waves) also arise naturally in some physical media (fluid and solid). A common example may be the tide waves in shallow water near a sea coast.

To explain the formation of discontinuities, we need to consider waves of finite amplitude and use nonlinear models (Logan, 1983). In contrast to the linear theory of wave propagation (where the wave shape is preserved), nonlinear phenomena involve a *distortion of the wave shape.* There are several physical mechanisms that lead to appearance of discontinuous waves. To illustrate how nonlinear effects produce discontinuous waves, let us consider a stress wave in a one-dimensional solid bar again. An initial wave caused by a force on one end of the bar may have a common (continuous) wave shape. Many materials have a property of transmitting disturbances at a speed that increases with increasing pressure (cause of the disturbance). Therefore, if the amplitude of the wave is not infinitesimal, then the disturbances of higher amplitude (higher points on the wave profile) move faster than ones of lower amplitudes. It means that the propagating wave will gradually become steeper until it turns into a discontinuous (shock) wave.

There are several nonlinear PDEs arising in various applications that take this "steeping" effect into account along with other physical phenomena: the Burger's equation, the Korteweg-deVries equation, the Boussinesq equation, the Sine-Gordon equation, the Born-Infeld equation, the nonlinear Schrodinger equation, see (Whitham, 1974). Here we restrict ourselves with the simplest model PDE – the so-called *nonlinear advective equation* :

$$u_t + c(u)\, u_x = 0, \qquad -\infty<x<\infty, \quad t>0, \tag{2.49}$$

In fact, the equation (2.49) is *quasi-linear*, i.e. linear in the derivatives u_t and u_x. The linear version of (2.49) with the constant parameter c is the simplest wave equation

$$u_t + cu_x = 0, \quad -\infty<x<\infty, \quad t>0. \tag{2.50}$$

It is easy to verify by direct substitution that the general solution of the linear equation (2.50) is the right-travelling wave $u(x,t) = f(x - ct)$ of an unchanged shape.

The nonlinear equation (2.49) may be derived directly from the differential conservation laws for mass and momentum. The corresponding technique was first developed by Riemann (the *Riemann's method*) applied to compression waves of finite amplitude in gases.

The nonlinear equation (2.49) simulates exclusively the effect of wave propagation with the speed $c(u)$ dependent on the displacement size u. Let us suppose that $c'(u)>0$, then larger values of the displacement $u(x,t)$ propagate faster. Then, as shown in (Logan, 1984), a discontinuity in the wave profile tends to form and shock waves will occur after a limited time of propagation.

Following our general idea, we outline one useful investigation technique for the nonlinear equation (2.49) based on nonlinear conservation laws.

2.4.1 Nonlinear Conservation Laws.

Quasi-linear PDEs similar to (2.49) are usually treated by the *method of characteristics* (Lin and Segel, 1974), which we will not discuss in detail here. Applied to a one-dimensional case, this method produces a family of curves in the (x,t)-plane called *characteristics* along which waves are propagated and along which the initial PDE is reduced to an ordinary differential equation. When the characteristics intersect, smooth solutions of the PDE (2.49) break down and a discontinuous solution is then developed and propagates as a shock wave. To determine the behaviour of the discontinuity, integral conservation laws may be used.

Namely, the equation (2.49) itself may be considered as a one-dimensional nonlinear conservation law (Logan, 1987). If we integrate (2.49) from $x=a$ to $x=b$, we obtain:

$$\frac{d}{dt}\int_a^b u(x,t)dx = F(u(b,t)) - F(u(a,t)), \tag{2.51}$$

where the function

$$F(u) = \int_0^u c(\theta)d\theta \tag{2.52}$$

is considered as a nonlinear flux. The equation (2.51) is the integral form of the same conservation law that the differential equation (2.50) represents. The differential equation (2.49) can be written via the function $F(u)$ as:

$$\partial u/\partial t + \partial F(u)/\partial x = 0.$$

The integral conservation law (2.51) is useful for finding the conditions across the discontinuity surface (which is developed after the characteristics intersect). Let $x=s(t)$ be the equation of the discontinuity surface and $a<s(t)<b$, then (2.51) leads to:

$$F(u_1) - F(u_0) = \frac{d}{dt}\int_a^{x(t)} u(x,t)dx + \frac{d}{dt}\int_{x(t)}^b u(x,t)dx =$$

$$= \int_a^{x(t)} u_t(x,t)dx + u_1\frac{ds}{dt} - \int_{x(t)}^b u_t(x,t)dx - u_0\frac{ds}{dt}.$$

Putting $a \to s(t)-0$ and $a \to s(t)+0$ in this formula, we obtain the following jump condition (in the notation (2.36)):

$$[F(u)] = [u]\, dx/dt. \tag{2.53}$$

It is easy to see the illustrated technique is similar to the one used for investigating discontinuities in the linearized model of solid bar dynamics in Section 2.3. Similarly, the obtained condition may be used for determination of the speed dx/dt of propagation of the discontinuity.

This technique may be also applied directly to mass and momentum conservation equations (2.6) and (2.11) to produce relations between sought-for functions on both sides of the discontinuity surface. For gas dynamics, such relations (the *Rankine-Hugoniot jump conditions*) are obtained from the conservation equations in (Logan, 1987).

Note that in some cases the nonlinear conservation law (2.49) might have a more complicated character. An interesting example is illustrated in (Svobodny, 1998). The equation (2.49) arises in the *technological process of ion etching* used in the manufacturing of semiconductor devices. This technology the semiconductor masks with a photoresistor and then bombards it with an ion beam. In this application, the sought-for function in the equation (2.49) is the slope $u(x,t)=\partial y(x,t)/\partial x$ of the surface $y(x,t)$, and the known function $F(u)$ in the above form of (2.49) represents the rate at which the surface is eroded by the ion beam. The flux function $F(u)$ appears to be *nonconvex*, even, with a local minimum at 0 and two or more maxima.

2.4.2 Impact of Diffusion and Dispersion.

So, waves of a finite amplitude can distort because the speed of propagation depends on the amplitude of the wave. As we said before, the equation (2.49) is a model equation that simulates only the nonlinear "shocking up effect" caused in the equation by the term $c(u)u_x$.

In real situations, there are also physical effects that tend to smooth the waves out. Two such basic phenomena are:

a) *diffusion* when a physical quantity flows from high concentrations to lower concentrations;
b) *dispersion* when the speed of wave propagation depends on the wavelength of the wave.

Both diffusion and dispersion may be linear or nonlinear. The *diffusion* phenomenon is most common and leads to smoothing a local distribution of a conserved quantity. It is explained in detail in Sections 3.3 and 3.4 for three-dimensional continuum media. Here we would like to mention that adding a diffusion term νu_{xx} to the model advection equation (2.49) at $c(u)=u$ leads to the well-known *Burger's equation* :

$$u_t + uu_x - \nu u_{xx} = 0, \qquad \nu>0, \tag{2.54}$$

whose travelling wave solution has a profile represented by a logistic curve and the wave profile is sharper for lower values of the diffusion parameter ν. In the case of a shock wave propagation, the diffusion term may represent a viscosity effect.

Another releasing phenomenon is *distortion* when the waves of various wavelength (or wave number) propagate at different speed. This effect is inherent in many physical processes (see, for example, the transverse vibrations of a beam in Chapter 3). Other examples include water waves and electromagnetic waves. Adding a dispersion term ku_{xxx} to the model advection equation (2.49) leads to the famous *Korteweg-deVries equation* :

$$u_t + uu_x + vku_{xxx} = 0, \qquad k>0, \tag{2.55}$$

whose travelling *solitary wave* solutions play a fundamental role in many applications.

Nonlinear wave dynamics is an exciting science but we shall not pursue it further. An interested reader can find more detail about the nonlinear wave phenomena in (Whitham, 1974).

2.5 Models of Viscoelasticity

Many real materials do not behave as elastic bodies. The theory of viscoelasticity (materials with memory) is a major source of applied integral equations (Corduneanu, 1991). Unlike the elasticity theory where the Hook's law is acting (see Section 2.1), the stresses in a viscoelastic body at a given moment depend on the entire prehistory of the deformation. In a one-dimensional homogeneous medium, it is described mathematically as

$$\sigma(x,t) = \int_{-\infty}^{t} G(t,s)\varsigma(x,s)ds, \tag{2.56}$$

where as in Section 2.1 $\sigma(x,t)$ denotes the stress and $\zeta(x,t)$ denotes the strain (deformation). The function $G(x,s)$ is called *the influence function* or the *Green's function.* For a stable medium (Maz'ya and Nikol'skii, 1989), the function $G(x,s)$ depends on the difference x-s.

The first integral model of elastic persistence in the form (2.56) was introduced by Boltzmann in XIX century (Boltzmann, 1874). Vito Volterra developed the Boltzmann theory and introduced the after-effect concept into other applications, in particular, into population ecology (Volterra, 1930). We will consider integral dynamical models in Chapter 4 and integral population models with after-effect in Chapter 5.

The following transformation generalizes the derivation of the model (2.25) of solid bar dynamics for the case when the Hook's law is replaced by the integral relation (2.56). Following (Volterra, 1930), we assume that the whole past history of the phenomenon is known:

$$\sigma(x,s) = \sigma_0(x,s), \quad -\infty < s \le 0.$$

From the momentum conservation equation (2.12), instead of (2.24) we obtain the following constitutive relation (see also (Nohel, 1982)):

$$\sigma(x,t) = \int_0^t G(t,s)\frac{\partial u(x,s)/\partial x}{1-\partial u(x,s)/\partial x}ds + \int_{-\infty}^0 G(t,s)\varsigma_0(x,s)ds$$

between the sought-for displacement $u(x,t)$ (in spatial coordinates) and stress component $\sigma(x,t)$. Substituting the last formula into the momentum conservation equation (2.12), we obtain

$$\rho\frac{\partial^2 u(x,t)}{\partial t^2} = \rho\tilde{f} + \int_0^t G(t,s)\frac{\partial}{\partial x}\frac{\partial u(x,s)/\partial x}{1-\partial u(x,s)/\partial x}ds, \tag{2.57}$$

$$\text{where } \tilde{f}(x,t) = f(x,t) + \int_{-\infty}^0 G(t,s)\varsigma_0(x,s)ds,$$

with respect to $u(x,t)$ and $\rho(x,t)$. The nonlinear integral-differential equation (2.58) describes the motion of the viscoelastic medium. The linearization technique for small displacements $u(x,t)$ reduces this equation to one linear integral-differential equation

$$\rho(x)\frac{\partial^2 u(x,t)}{\partial t^2} = \rho(x)\tilde{f}(x,t) + \int_0^t G(t,s)\frac{\partial}{\partial x}u(x,s)ds, \tag{2.58}$$

with respect to $u(x,t)$. Here $\rho(x)$ is the given initial density.

The elastic heredity may be more complex than described by the Boltzman formula (2.56). Let us consider the nonlinear relationship between stress and strain:

$$\sigma(x,t) = \Phi(\varsigma(x,t)) + \int_{-\infty}^t G(x,t,s)\Psi(\varsigma(x,s))ds,$$

Then the analogue of the motion equation (2.58) (for small displacements u) is the following nonlinear integrodifferential equation with respect to the sought-for displacement $u(x,t)$:

$$\rho(x)\frac{\partial^2 u(x,t)}{\partial t^2} = \frac{\partial}{\partial x}\Phi\left(\frac{\partial u(x,s)}{\partial x}\right) + \frac{\partial}{\partial x}\int_0^t G(x,t,s)\frac{\partial u(x,s)}{\partial x}ds + \\ + \rho(x)\tilde{\tilde{f}}(x,t) \tag{2.59}$$

There are many variations of viscoelastic behaviour. For example, the linear elastic heredity of a *creep type* is described by the following relation (Arutjunian and Kolmanovskii, 1983):

$$\varsigma(x,t) = J(0)\left[\sigma(x,t) + \int_{-\infty}^t g(t-s)\sigma(x,s)ds\right],$$

where

$J(t)$ is the *creep compliance* function and

$g(t)=[J(0)]^{-1}dJ(t)/dt$.

Then the corresponding linear equation of motion is (Corduneanu, 1991) :

$$\frac{\partial^2 u(x,t)}{\partial t^2} - c^2\frac{\partial^2 u(x,s)}{\partial x^2} - \int_0^t \bar{g}(t-s)\frac{\partial^2 u(x,s)}{\partial x^2}ds = \bar{f}(x,t), \tag{2.60}$$

where $c^2=[\rho J(0)]^{-1}$,

$\bar{g}(t)$ is the inverse operator to $g(t)$,

$\bar{f}(x,t)$ is a given function that includes the prehistory of the process.

Further references about the mathematical theory of viscoelastic materials may be found in (Hrusa, 1984; Renardy et al, 1987; Leugering, 1987; Nohel, 1982) and other publications.

3. THREE-DIMENSIONAL CONSERVATION LAWS AND MODELS

Here we briefly expose general three-dimensional conservation (balance) laws and some classical equations of solid and fluid dynamics that give a basis for some applied three-dimensional models of Section 4. An interested reader can find more details about three-dimensional conservation laws in (Lin and Segel, 1974).

This section provides the material necessary to derive the equations of motion in rather general cases but illustrates complete models in the simplest cases only (perfect fluids and gases, homogeneous elastic bodies). More complex applied situations can be also treated on the basis of given conservation laws and constitutive relations of continuum mechanics, such as non-homogeneous physical media, nonlinearities, multi-phase media, non-steady initial states, disconnected regions and moving boundaries, interaction of various external factors and forces, and so on.

3.1 Mass Conservation and Continuity Equations

The flux (the flow of material) across the boundary surface $S=\partial R$ of a volume R is equal to

$$\iint_{\partial R} \rho \mathbf{v} ds = \iiint_R \operatorname{div}(\rho \mathbf{v}) dx_1 dx_2 dx_3,$$

where the right-hand part is obtained by the Gauss divergence theorem (see Appendix).

Here $\mathbf{v}=(v_1,v_2,v_3)$ is the velocity and $\rho(\mathbf{x},t)$ is the mass density per unit volume at point $\mathbf{x}=(x_1,x_2,x_3)$ at time t. Since mass is neither created nor destroyed (no sources and sinks inside the volume), then

$$\iiint_R \left\{ \frac{\partial \rho}{\partial t} + \operatorname{div}(\rho \mathbf{v}) \right\} dx_1 dx_2 dx_3 = 0 \tag{2.61}$$

for a fixed space region R. The formula (2.61) represents *the integral form of the mass balance (conservation law)*.

Since (2.61) is to be true for all arbitrary volume elements, then by Dubois-Reymond lemma (see Appendix) we get the differential form of the mass balance:

$$\frac{\partial \rho}{\partial t} + \nabla(\rho \mathbf{v}) = 0 \tag{2.62}$$

or

$$\frac{\partial \rho}{\partial t}+\frac{\partial}{\partial x_1}(\rho v_1)+\frac{\partial}{\partial x_2}(\rho v_2)+\frac{\partial}{\partial x_3}(\rho v_3)=0,$$

which is commonly known as the *continuity equation* (Kapur, 1988).

If the medium (fluid) is incompressible, then $\rho(\mathbf{x},t)=\rho$ =const and the continuity equation becomes

$$\operatorname{div}(\mathbf{v})=0 \quad \text{or} \quad \frac{\partial v_1}{\partial x_1}+\frac{\partial v_2}{\partial x_2}+\frac{\partial v_3}{\partial x_3}=0.$$

If the *rotation* of the velocity $\mathbf{v} = \nabla \times \mathbf{v}$ = rot Ψ is equal to 0 (the rotation is defined in Section 1.1 of Appendix), then the motion (flow) is called *irrotational* and a scalar velocity *potential function* $\Phi=\Phi(\mathbf{x},t)$ exists such that

$$\mathrm{v}=-\operatorname{grad}(\Phi) \quad \text{or} \quad v_1=\frac{\partial \Phi}{\partial x_1},\quad v_2=\frac{\partial \Phi}{\partial x_2},\quad v_3=\frac{\partial \Phi}{\partial x_3},$$

Then the continuity equation (2.62) is reduced to the *Laplace equation*:

$$\nabla^2 \Phi = 0 \qquad (\text{or} \quad \Delta \Phi = 0).$$

3.2 Momentum Conservation and Cauchy Equations

As in the one-dimensional case (see Section 2.2), the momentum balance generalizes the **Newton's second law of mechanics** that states that the rate of a change of momentum is equal to total force. The momentum of a three-dimensional material region Ω is $\iiint_\Omega \rho \mathbf{v}\, dx_1 dx_2 dx_3$.

The forces acting on the region are of two types: *body forces* denoted by $\mathbf{f}(x,t)$ and *surface forces* represented by the *stress vector* $\sigma(x,t,\mathbf{n})$ where $\mathbf{n}$ is the outward normal to the surface. The following stress principle was postulated by Cauchy

$$\frac{d}{dt}\iiint_R \rho \mathbf{v}\, dx_1 dx_2 dx_3 = \iiint_R \rho \mathbf{f}\, dx_1 dx_2 dx_3 + \iint_{\partial R} \boldsymbol{\sigma}\, ds \tag{2.63}$$

Instead of a single scalar *stress component* $\sigma(x,t)$ in one-dimensional case, we need to introduce a 3×3 matrix called (scalar) *stress tensor* $\sigma(\mathbf{x},t)=\{\sigma_{ij}\}_{ij=1,2,3}$ that defines the dependence of the stress vector $\boldsymbol{\sigma}$ on the unit normal $\mathbf{n}$:

$$\boldsymbol{\sigma}(\mathbf{x},t,\mathbf{n}) = \mathbf{n} \cdot \sigma(\mathbf{x},t)$$

The nine functions $\sigma_{ij}(\mathbf{x},t)$, $i,j=1,2,3$, are called the *components of the stress tensor* σ.

After some transformation, we obtain the following *integral form of the momentum conservation law* :

$$\iiint_R \left\{ \rho \frac{Dv_j}{Dt} - \rho f_j - \sum_{i=1}^{3} \frac{\partial \sigma_{ij}}{\partial x_i} \right\} dx_1 dx_2 dx_3 = 0, \;\; j = 1,2,3. \qquad (2.64)$$

Assuming that the intergrand in (2.64) is continuous, we can write the *differential form of the momentum conservation law* :

$$\rho \frac{Dv_j}{Dt} = \rho f_j + \sum_{i=1}^{3} \frac{\partial \sigma_{ij}}{\partial x_i}, \qquad j = 1,2,3. \qquad (2.65)$$

The equations (2.65) are known as *Cauchy equations*. Along with the continuity equation (2.62) they connect 13 sought-for variables: density ρ, three components of the velocity v, and 9 components of the stress tensor σ. To close the model of a process, we need to consider more specific situations and use various *constitutive relations* that are valid in the specific case. Some common examples are mentioned below.

3.2.1 Conservation of Angular Momentum

As opposed to a one-dimensional case, in space we have one more momentum conservation law - for *angular momentum* (of the momentum of momentum $\rho\mathbf{v}$) $\mathbf{x} \times \rho\mathbf{v}$ about the origin of coordinates (see Lin and Segel, 1973):

$$\frac{d}{dt} \iiint_R dx_1 dx_2 dx_3 = \iiint_R (\mathbf{x} \times \rho\mathbf{f}) dx_1 dx_2 dx_3 + \iint_{\partial R} (\mathbf{x} \times \rho\boldsymbol{\sigma}) ds.$$

It can be proven that this conservation law is equivalent to the symmetry property of the stress tensor:

$$\sigma_{ij}(\mathbf{x},t) = \sigma_{ji}(\mathbf{x},t), \quad i,j=1,2,3.$$

Then we have only six unknown stress components (instead of nine in the general case). The law of angular momentum conservation is not true for any medium, for example, for some anisotropic solids and polar fluids.

3.2.2 Newtonian Viscous Fluids

Viscous fluids with certain additional assumptions about symmetry and invariance of the stress tensor are called *Newtonian fluids*. For such fluids, the stress tensor $\sigma_{ij}(\mathbf{x},t)$ has the form (*the Stokes law*):

$$\sigma_{ij} = (-p + \lambda\, \nabla\mathbf{v})\delta_{ij} + \mu\,(\partial v_i/\partial x_j + \partial v_i/\partial x_j), \quad i,j=1,2,3.$$

where λ and μ are *coefficients of viscosity* and δ_{ij} is the Kroneker symbol (Maz'ya and Nikol'skii, 1989). Then the Cauchy equations (2.65) have the form:

$$\rho\frac{Dv_j}{Dt} = \rho f_j - \frac{\partial}{\partial x_i}(p - (\lambda + \mu)\nabla\,\mathbf{v}) + \mu\nabla^2 v_j, \quad j = 1,2,3.$$

and are known as the *Navier-Stokes equations*.

Using the definition of the complete derivative $D/Dt = \partial/\partial t + \mathbf{x}\cdot\nabla$, the Navier-Stokes equations may be written as following:

$$\frac{\partial \mathbf{v}}{\partial t} + (\mathbf{v}\cdot\nabla)\,\mathbf{v} = -\rho^{-1}\nabla p + (\lambda + \mu)\nabla\,\mathbf{v} + \mu\nabla^2\mathbf{v} + \mathbf{f}. \qquad (2.66)$$

For incompressible Newtonian fluids the mass conservation law (the continuity equation) appears to be

$$\operatorname{div}\mathbf{v} = 0$$

and the Navier-Stokes equations (2.66) are:

$$\frac{\partial \mathbf{v}}{\partial t} + (\mathbf{v} \cdot \nabla)\, \mathbf{v} = -\rho^{-1}\nabla p + \mu\nabla^2 \mathbf{v} + \mathbf{f}.$$

The first term on the left in (2.66) disappears in the case of a steady flow. In the case of slow flows, we can linearize the Navier-Stokes equations and omit the nonlinear term $(\mathbf{v}\cdot\nabla\mathbf{v})$ that is small as compared with the remaining terms. This procedure leads to the *Stokes equations*:

$$\mu\nabla^2\mathbf{v} = \rho^{--1}\nabla p + \mathbf{f}, \qquad \text{div } \mathbf{v} = 0,$$

which describe the *slow steady flow of a viscous fluid.* If there is no external force: $\mathbf{f} = 0$, then we obtain *homogenous Stokes equations*:

$$\mu\nabla^2\mathbf{v} - \rho^{--1}\nabla p = 0, \qquad \text{div } \mathbf{v} = 0.$$

This is a system of four equations with respect to four sought functions v_1, v_2, v_3 and p.

3.2.3 Inviscid Fluids

For inviscid fluids, the stress $\sigma_{ij}(\mathbf{x},t)$ is proportional to the normal $\mathbf{n}$:

$$\sigma(\mathbf{x},t) = -p(\mathbf{x},t)\mathbf{n},$$

where the coefficient of proportionality $p(\mathbf{x},t)$ is the *pressure*. Then the Cauchy equations (2.65) are transformed into the following *Euler equations*:

$$\rho\frac{Dv_j}{Dt} = \rho f_j - \frac{\partial p}{\partial x_i}, \quad j = 1,2,3,$$

or

$$\frac{\partial v_1}{\partial t} + v_1 \frac{\partial v_1}{\partial x_1} + v_2 \frac{\partial v_1}{\partial x_2} + v_3 \frac{\partial v_1}{\partial x_3} = f_1 - \frac{1}{\rho} \frac{\partial p}{\partial x_1},$$
$$\frac{\partial v_2}{\partial t} + v_1 \frac{\partial v_2}{\partial x_1} + v_2 \frac{\partial v_2}{\partial x_2} + v_3 \frac{\partial v_2}{\partial x_3} = f_2 - \frac{1}{\rho} \frac{\partial p}{\partial x_2}, \qquad (2.67)$$
$$\frac{\partial v_3}{\partial t} + v_1 \frac{\partial v_3}{\partial x_1} + v_2 \frac{\partial v_3}{\partial x_2} + v_3 \frac{\partial v_3}{\partial x_3} = f_3 - \frac{1}{\rho} \frac{\partial p}{\partial x_3}.$$

Three equations (2.67) and the continuity equation (2.62) connect five sought-for variables: density ρ, three components of the velocity v, and the pressure p. To find a complete model of the process, an *equation of state* (like energy conservation relations) is usually used to reveal the specific properties of the fluid.

The simplest state equation is the a so-called *barotropic equation* $p=F(\rho)$, in which the pressure depends only on the density. It allows us to eliminate the pressure p from the Euler equations (2.67). For a general fluid motion, it is not valid and the equation of state usually introduces an additional unknown variable (temperature, energy, entropy).

3.2.4 Propagation of Sound in Space

The *acoustic approximation* is based on a linearization of the equations of continuity and momentum under the assumption that disturbances of the unknown functions (the pressure p and density ρ) from a constant equilibrium state ρ_0, p_0 are small:

$$\rho = \rho_0(1 + s), \qquad p = p_0 + c^2\rho_0 s.$$

For small disturbances s, the equations (2.62) and (2.67) give the following equations of gas motion (Kapur, 1988):

$$\rho_0 \partial \mathbf{v}/\partial t = -c^2 \rho_0 \nabla s, \quad \rho_0 \partial s/\partial t + \rho_0 \nabla \mathbf{v} = 0.$$

If the motion is irrotational (see Section 3.1), then $\mathbf{v}=-\mathrm{grad}\,\Phi$ for a scalar potential function $\Phi=\Phi(\mathbf{x},t)$ and the last equations are transformed into the *three-dimensional wave equation*:

$$\Phi_{tt} = c_0^2 \nabla^2 \Phi,$$

where the parameter c_0 is the *speed of acoustic waves propagation.*

The propagation of harmonic acoustic waves is considered in more detail in Sections 1.3 and 2 of Chapter 4.

3.2.5 Elastic Waves in Solids

Let us consider small movements of perfect, homogeneous, isotropic elastic solids. As in the one-dimensional case (Section 2.1), additional constitutional relations for solid dynamics should be derived from the elasticity theory.

An elastic body under deformation is described by the displacement vector $\mathbf{u}(\mathbf{x}) = \{u_1, u_2, u_3\}$ and the stress tensor $\sigma_{ij}(\mathbf{x},t)$, i,j=1,2,3. We suppose that the angular momentum conservation law holds true, and therefore, the stress tensor is symmetric: $\sigma_{ij} = \sigma_{ji}$.

To describe the deformation of solid bodies, we need to establish state equations that relate the stress to the kinematical characteristics of deformation (similar to the relation used in Section 3.2.3). The latter characteristics in the elasticity theory are described by the *strain tensor*:

$$\{D_{jj}\} = 1/2\{\partial v_i/\partial x_j + \partial v_i/\partial x_j\}, \quad i,j=1,2,3.$$

If we accept linear relationships between the components of the stress and strain tensors, then in the case of a homogenous isotropic body the following equations are true (*the three-dimensional Hook's law*):

$$\sigma_{ij} = \lambda\delta_{ij}\, \nabla\mathbf{u} + \mu\,(\partial u_i/\partial x_j + \partial u_i/\partial x_j), \qquad i,j=1,2,3.$$

where the parameters λ, μ are physical constants of the medium called the *Lame's constants.* The Hook's law is true for the *small* deformations only.

Substitution of the Hook's law into the linear momentum conservation law (2.65) (Cauchy equations) and its linearization delivers the following *linearized* equation of motion :

$$\rho\, \partial^2\mathbf{u}/\partial t^2 - (\lambda+2\mu)\nabla(\nabla\mathbf{u}) + \mu\nabla\times\nabla\times\mathbf{u} = \mathbf{f}.$$

Without any restriction, the sought-for displacement vector **u** may be represented as the composition of its rotational and its irrotational components:

$$\mathbf{u} = \mathbf{u}_P + \mathbf{u}_S, \quad \mathbf{u}_P = \nabla \Phi = \mathrm{grad}\,\Phi, \quad \mathbf{u}_S = \nabla\times\Psi = \mathrm{rot}\,\Psi$$

where the functions Φ and Ψ are the scalar and vector potentials (Filippi et al, 1999) and the rotation rot Ψ is defined in Section 1.1 of Appendix. Similarly, the given external force **f** may be represented as

$$\mathbf{f} = \mathbf{f}_P + \mathbf{f}_S, \quad \mathbf{f}_P = \nabla\, G, \quad \mathbf{f}_S = \nabla\times\mathbf{H}.$$

Then the linearized motion equation is decomposed into two *wave equations* with respect to functions $\mathbf{u}_P$ and $\mathbf{u}_S$:

$$\partial^2\mathbf{u}_P/\partial t^2 - c_1^{\,2}\nabla^2\mathbf{u}_P = c_1^{\,2}\,(\lambda+2\mu)^{-1}\nabla G,$$

$$\partial^2\mathbf{u}_S/\partial t^2 - c_2^{\,2}\nabla^2\mathbf{u}_S = c_2^{\,2}\mu^{-1}\nabla\times\mathbf{H},$$

where $c_1^{\,2}=(\lambda+2\mu)/\rho,\ c_2^{\,2}=\mu/\rho.$

The potentials Φ and Ψ also satisfy some wave equations:

$$\partial^2\Phi/\partial t^2 - c_1^{\,2}\nabla^2\Phi = c_1^{\,2}\,(\lambda+2\mu)^{-1}\nabla G,$$

$$\partial^2\Psi/\partial t^2 - c_2^{\,2}\nabla^2\Psi = c_2^{\,2}\mu^{-1}\nabla\times\mathbf{H}.$$

So, in dynamics, the wave equations for elastic stress waves are valid in presence of body forces. Let us consider the equilibrium case. Substituting $\partial^2\mathbf{u}/\partial t{=}0$ into the linearized equation of motion, one can obtain the following *Lame's equation*:

$$(\lambda+2\mu)\nabla(\nabla\,\mathbf{u}) = \mathbf{f},$$

which is an analogue of the Poisson equation (and Laplace equation at **f**=0).

The results obtained above show the existence of two different types of waves with different propagation speeds c_1 and c_2:

- *compression/expansion waves* (or *pressure waves*) that propagate at the speed c_1;
- *shear/distortion waves* that propagate at the speed c_2.

In a two-dimensional case, the pressure waves are *longitudinal* waves and describe the vibrations parallel to the direction of propagation. The shear waves are *transverse* waves that describe the vibrations perpendicular to the direction of propagation.

In seismology the pressure waves are also known as *primary waves* or *P-waves* and shear waves as *secondary waves* or *S-waves* because the latter propagate at lower speed and arrive later then the former (very roughly $c_2=\frac{1}{2}c_1$). In applied engineering problems, the importance of these two wave types may be different. While common investigation techniques consider these waves separately, the interaction between them might be essential at discontinuities (interfaces).

3.3 Energy Balance and Thermodynamics

In a three-dimensional continuum medium (solid or fluid), the distributed thermodynamic state variables (specific pressure p, specific absolute temperature θ, specific internal energy e, specific entropy s) depend on the spatial coordinate $\mathbf{x}$ and time t.

As in the one-dimensional case, we will postulate the balance of energy law in integral form. It states that the change of the total energy equals the total work that all the forces (body forces $\mathbf{f}$ and surface forces σ) do on the region plus the heat flux $\mathbf{f}$ into the region:

$$\frac{d}{dt}\iiint_R\left(\frac{1}{2}\mathbf{v}\cdot\mathbf{v}+e\right)\rho d\mathbf{x}=\iiint_R(\mathbf{f}\cdot\mathbf{v})\rho d\mathbf{x}+\iint_{\partial R}(\boldsymbol{\sigma}\cdot\mathbf{v}-\mathbf{q}\cdot\mathbf{n})ds.$$

Providing standard reasoning as for the above integral conservation laws, one can obtain the following *balance of energy* in the integral form from the previous formula (Lin and Segel, 1974):

$$\iiint_R\rho\frac{D}{Dt}\left(\frac{1}{2}\mathbf{v}\cdot\mathbf{v}+e\right)d\mathbf{x}=\iiint_R[\rho(\mathbf{f}\cdot\mathbf{v})-\nabla\cdot(\mathbf{q}-\boldsymbol{\sigma}\cdot\mathbf{v})]d\mathbf{x}. \qquad (2.68)$$

Its differential analogue is :

$$\rho\frac{D}{Dt}\left(\frac{1}{2}\mathbf{v}\cdot\mathbf{v}+e\right)=\rho(\mathbf{f}\cdot\mathbf{v})-\nabla\cdot(\mathbf{q}-\boldsymbol{\sigma}\cdot\mathbf{v}),\qquad(2.69)$$

where $\boldsymbol{\sigma}=\{\sigma_{ij}\}$ is the 3×3-matrix function of the *stress tensor* (see Section 3.2).

These equations are supplemented with the same Gibbs relation (2.15) as in the one-dimensional case followed from the first and second laws of thermodynamics:

$$\frac{De}{Dt}=\theta\frac{Ds}{Dt}-p\frac{Ds}{Dt}(1/\rho).\qquad(2.70)$$

The energy balance (2.68) is used in the next section as a basic conservation law for description of heat diffusion.

3.4 Heat Balance and Diffusion Processes

The diffusion (from Latin word *diffusio*) is a fundamental physical phenomenon. It reflects the tendency for relatively high concentrations of a specific physical quantity to disperse, i.e. flow from high concentrations to lower concentrations. The physical mechanism that implements this phenomenon works on the molecular level and reflects the so-called *random walk of molecules*(see also Chapter 5). So, the diffusion is a macro description of molecular processes.

The diffusion flux represents a "total force" that averages (smoothes) a local distribution of a conserved quantity. The diffusion flux across the boundary surface $S=\partial R$ of a volume R is equal to $\iint_S \mathbf{q}\,ds$.

Two main examples of the physical quantities undergo to diffusion are the *heat* and *the concentration of a chemical substance* in a continuum medium (gas or fluid).

As it follows from **the second law of thermodynamics**, the heat flows from a body with higher temperature to a body with lower temperature. The heat conduction is governed by two different physical laws:

– the energy conservation law (that is an example of fundamental law);
– the Fourie's heat law (an example of empirical constitutive relation).

Fourie's heat law states that the heat flux density (the rate of heat flow per unit area) is proportional to the temperature gradient $\nabla\theta$:

$$\mathbf{q} = -K\mathrm{grad}\,\theta = -K\nabla\theta, \tag{2.71}$$

where
θ is the absolute temperature of the medium,
K is a material parameter known as the *thermal conductivity* (usually depends on θ).

If we consider the *concentration of particles* immersed in a solvent than the empirical relation (2.71) still holds and describes the diffusion of the solute. In this case, the diffusion means the fundamental process of a penetration of molecules of one substance into another substance during their contact, due to the heat motion of the molecules. The diffusion leads to a spontaneous equilibration of the substance concentration in the space. In this case, the formula (2.71) is called **Fick's law** and the parameter K is the *diffusivity* of the substance.

Let $e(\mathbf{x},t)$ be the energy density per unit volume at point $\mathbf{x}=(x_1,x_2,x_3)$ at time t. We assume that the heat may be produced or consumed inside the region R, say, by chemical reactions. The conservation of the total energy takes the form

$$\frac{dE}{dt} = \frac{d}{dt}\iiint_R e(\mathrm{x},t)d\mathbf{x} = -\iint_{\partial R} \mathrm{q}ds + \iiint_R f(\mathrm{x},t)d\mathbf{x} \tag{2.72}$$

for a fixed space region R, where the corresponding *heat source function* $p(\mathbf{x},t)$ describes the heat generated (consumed) per volume per unit time in the region R.

The heat energy increase ΔE due to the temperature increase $\Delta\theta$ is found to be:

$$\Delta E = C\Delta\theta,$$

where C is a material parameter known as the *heat capacity*, which also depends on θ. Then the energy density

$$e = c\rho\theta,$$

where ρ is the mass density and the *specific heat* c is a the heat capacity per unit mass (may depend on θ).

Using the last formula, divergence theorem and (2.72), we obtain the following *integral form of the heat balance law*:

$$\iiint_R \left\{ \frac{\partial(c\rho\theta)}{\partial t} - \operatorname{div}(K\nabla\theta) - f \right\} d\mathbf{x} = 0. \tag{2.73}$$

3.4.1 Diffusion Equation

Assuming smoothness of the integrand in (2.73), by Dubois-Reymond lemma we obtain the differential form of the heat balance:

$$\frac{\partial(c\rho\theta)}{\partial t} = \operatorname{div}(K\nabla\theta) + f, \tag{2.74}$$

which is commonly known as the *heat conduction equation.*

The equation (2.74) is nonlinear because the thermal conductivity K depends on θ. The specific density ρ and heat c can be usually considered to be constants, then the heat equation (2.74) may be written as:

$$\frac{\partial\theta}{\partial t} = \operatorname{div}(k\nabla\theta) + \frac{f}{c\rho}, \qquad k = \frac{K}{c\rho}, \tag{2.75}$$

or

$$\frac{\partial\theta}{\partial t} = \frac{\partial}{\partial x_1}\left(k\frac{\partial\theta}{\partial x_1}\right) + \frac{\partial}{\partial x_2}\left(k\frac{\partial\theta}{\partial x_2}\right) + \frac{\partial}{\partial x_3}\left(k\frac{\partial\theta}{\partial x_3}\right) + \frac{f}{c\rho}.$$

If the thermal conductivity K can be also constant, then the corresponding linear equation (2.75) is called the *nonhomogeneous diffusion equation* :

$$\frac{\partial \theta}{\partial t} = k\nabla^2 \theta + \frac{p}{c\rho}. \tag{2.76}$$

If the source function $p=p(\mathbf{x},t)\equiv 0$, then the equation (2.76) is the *homogeneous diffusion equation.*

3.4.2 Advection-Diffusion Equation

Let us consider an advective-diffusive process where *solute particles* in a solvent are diffused and transported with the flow of the solvent. *Advection* (from Latin word *advectio*) means a horizontal transfer of a fluid or a gas together with its properties, such as humidity, heat, chemical substances, etc.

If the sought-for function $C(\mathbf{x},t)$ represents the concentration of the diffusing substance, then the governing equation of motion is still the integral conservation law (2.72) with the replaced quantity $e=C$. However, the constitutive relation (2.71) for the flux density $\mathbf{q}$ has the following form:

$$\mathbf{q} = -D\nabla C + C\mathbf{v}, \tag{2.77}$$

The formula (2.77) includes the diffusion term $D\nabla C$ (reflecting the Fick's law) and the advective term $C\mathbf{v}$, where vector $\mathbf{v}$ represents the speed of the solvent flow.

The formulae (2.72) and (2.77) produce the following *advection-diffusion equation*:

$$\partial C/\partial t + \text{div}(C\mathbf{v}) = \text{div}(D\nabla C) + f, \tag{2.78}$$

where the parameter D is the *diffusivity* of the substance (measured per unit volume and unit time) and the function $\mathbf{v}$ is the given *speed of advection.*

The external source function $f=f(\mathbf{x},t)$ in (2.78) depends on $\mathbf{x}$, t and may also depend on other endogenous model variables. In specific cases, the source function f can reflect such external forces and processes as gravity, adsorption, sedimentation, physical-chemical transformations of the substance, and so on (see applied models in the next section).

If the diffusivity D is constant and there are no external sources, f=0, and no advection, $\mathbf{v}$=0, then the equation (2.78) leads to the linear *homogenous diffusion equation*:

$$\partial C/\partial t = D\,\nabla^2 C. \qquad (2.79)$$

If there is no diffusion, D=0, and no external sources, f=0, then the equation (2.78) is reduced to the linear *advective equation*:

$$\partial C/\partial t + \mathrm{div}(C\mathbf{v}) = 0 \qquad (2.80)$$

A one-dimensional version (2.50) of the advective equation (2.80) was considered in Section 2.4.

4. APPLIED MODELLING OF WATER TRANSPORT AND CONTAMINATION

In the previous section, we have derived basic equations of three-dimensional continuum media (gas, fluids and solids). Deriving a set of the equations of motion (even in the case when the number of the equations matches the number of sought-for functions) does not yet provide a workable applied model of the process. There is still a lot of work to do in deciding which model relations are relevant and which are not, determining possible constraints and control influences, initial and boundary conditions, and so on. To proceed further with the modelling process, we need to choose a specific applied system (or process) and define main purposes of modelling such as prediction, control, optimization, and so on.

To illustrate the specifics of applied mathematical modelling, in this section we consider applied models of water transport and contamination. Such models are used for two major engineering activities:

- control of quality and preserving water resources;
- prediction and control of flooding water dissemination.

The goal of modelling is evaluation of the level of water contamination with some harmful substance and prediction of water distribution (a flood plan).

Simple models of water pollution propagation are constructed similarly to basic models of Section 3. In this section, we consider more complex models used in applied research, in particular, for forecasting radioactive contamination of Ukrainian rivers after the Chernobyl nuclear disaster.

4.1 Description of Physical Processes.

Forecast methods for pollution migration in rivers, lakes and artificial water reservoirs are based on a mathematical description of general hydrodynamic, hydraulic and physical-chemical processes controlling pollution transfer in these water objects. The modelling will take into account the following phenomena relevant for quality of the water environment:

Pollution (*contamination*) is introducing a harmful chemical, physical, biological or other substance (*pollutant ingredient, agent*) into the environment. There are more than 2000 known pollution substances with negative effects for the environment. The most common are the carbonic acid gas CO_2, the carbon monoxide CO, the sulphur oxide SO_2, the nitric oxide NO_2, the ammonia NH_3.

Sedimentation (from Latin word *sedimentum* - settling) is the process of deposit (sediment) of solid particles suspended in a fluid or a gas under the influence of the gravity force.

Stratification (from Latin words *stratum* - layer and *facere* - make) is a vertical distribution of the temperature in water medium or atmosphere in accordance with the height, which determines the intensity of vertical transfer. It causes vertically distributed water layers with different density that suppress the heat exchange and other physical processes in water reservoirs.

Temperature inversion is a situation in the air or water medium when its high layers (strata) have a higher temperature compared with the temperature of low layers. It leads to a disturbance of the vertical diffusion of pollution agents. In atmosphere, the warmer layer squeezes polluted air (a low layer) and the risk of *smog* is raised.

In general, the applied models of water pollution propagation need to take the following *hydrophysical processes* into account:

- wind and water flow currents;
- distribution and transformation of wind waves;
- riverside currents generated by waves;
- dynamics of water stratification;

- transport of suspended drifts;
- transport of involved drifts;
- sedimentation and bottom wash-out (disturbance of drifts).

Because a natural water reservoir is not a non-homogeneous medium (it has particles in suspension, changes of temperature, etc.), the density and the wave speed of the water are functions of spatial coordinates. In real-life situations, there may be obstacles of different sizes (from small particles to fish). The models also have to take into account the geometrical shape and physical characteristics of the bottom of water reservoirs. The models usually use the following input data (as initial and boundary conditions):

- hydrological conditions of the reservoir;
- morphological data (depth, bottom profile and so on);
- meteorological conditions;
- the location and intensity of contamination sources.

4.2 Classification of Models

Consideration of the hydrophysical phenomena mentioned above allows us to determine the intensity of transfer mechanisms and pollution accumulation in reservoirs. Correspondingly, applied models of pollution propagation in water objects contain several or all of the *following submodels* (*model blocks*):

- a model of water dynamics;
- a model of the transport of suspended particles;
- a model of the pollution transfer in solute;
- a model of the pollution transfer in suspense;
- a model of pollution accumulation in bottom sediments;
- a model of the pollution transfer in the bottom sediments involved in the erosion-sedimental processes;
- a model of pollution transfer in the inter-pore water of bottom sediments;
- models of physical and chemical transformation of the pollution agent.

A scheme of the interaction of these model blocks is given in the Fig. 2.1.

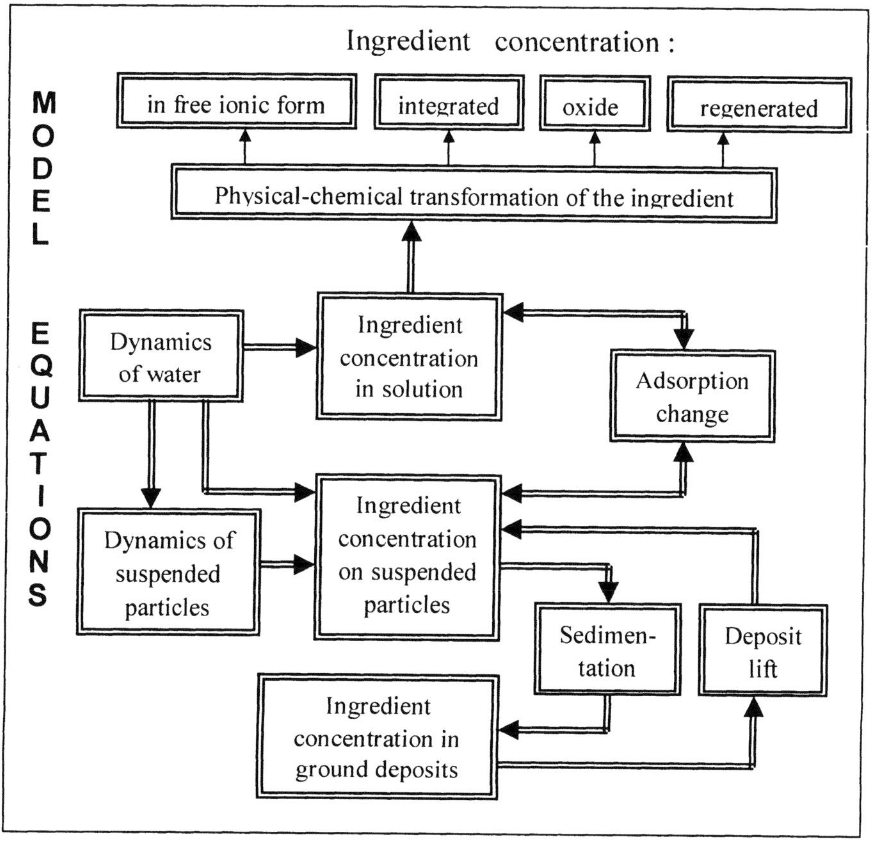

Figure 2.1. Block diagram of the applied models of water contamination

A representative set of applied models has been developed and used in practice for describing such complicated interacting processes. These models differ from each other by their dimension and the range of time space averaging of the corresponding physical processes. A classification of the models is given below.

1. Three-dimensional models give the most exhaustive description of the process under consideration. However, practical drawbacks of the three-dimensional model arise in getting specific detailed data required for the forecast (an initial state, boundary currents, bottom morphology, etc.).

Another difficulty lies in high computational complexity of the model. For these reasons such models are rarely used in practical calculations.

2. Two-dimensional plane models. In many real-life problems, an assumption may be admitted about a weak vertical mutability of the pollution transfer. When the phenomena under study have a space scale that exceeds several hundred meters, it is usually enough to compute averaged by depth characteristics of the water current without a detailed vertical distribution of the pollutant concentration.

3. Two-dimensional vertical models describe the vertical distribution of a pollutant concentration. They are used for investigation of the accumulation of the pollution agent in specific areas where the depth changes sharply, for instance, in river depression-traps.

4. One-dimensional river-bed models describe dynamics of water current parameters averaged through the river cross-section.

5. One-dimensional vertical models describe dynamics of current parameters averaged by a horizontal section.

6. Zero-dimensional models are based on the assumption that the pollution concentration is constant within some parts of the water reservoir (compartment, cell, chamber). Such models describe the dynamics of the pollutant exchange between two neighboring compartments and are based on ordinary differential equations. The term *"compartment* (*cell, chamber*) *models"* is usually applied to the zero-dimensional *models of full intermixing*, when the average pollution concentration in the stream flowing out from a considered compartment takes equal to the pollution concentration in the compartment. Sometimes, separation into compartments along the current depth is also used for stratified reservoirs.

The three- and two-dimensional models of water pollution propagation are considered in more detail below.

4.3 Three-Dimensional Model

Let us consider the process of propagation of a pollution substance in a domain of a water medium with the coordinates x_1, x_2, x_3. Corresponding three-dimensional mathematical model includes the following model blocks:

- the equations *of water dynamics* and *suspended drifts transport*;
- the equations *of* the *pollution concentrations in dissolute form C, in suspense* C^s and *bottom sediments* C^b;
- the equations *of sorption interchange, sedimentation and stirring up processes*.

The model appears to be too general for practical use but it provides a proper theoretical basis for simpler models exposed further.

4.3.1 Equation of Transport of the Ingredient in Solute

The equation of advective-diffusive transport for the concentration $C(x_1,x_2,x_3,t)$ of the ingredient (a harmful substance) in dissolved phase is of the form:

$$\partial C/\partial t + \sum_{i=1}^{3} \partial(v_i C)/\partial x_i = \sum_{i=1}^{3} \partial(A_i \partial C/\partial x_i)/\partial x_i - \\ - \alpha_{12}(K_d C - C^s)), \qquad (2.81)$$

where

v_i are the components of the water flow velocity,

A_i are the diffusion coefficients, i=1,2,3,

the parameter α_{12} characterises the intensity of adsorption exchange in the system "water-suspension".

This equation is a generalization of the advection-diffusion equation (2.78) that takes the adsorption exchange process into account, see also section 4.3.5.

Commonly used *boundary conditions* for the equation (2.81) are of the form:

(a) on the free water surface $x_3=\eta$:

$$A_3 \partial C/\partial x_3 = - v_3 C, \qquad (2.82)$$

(b) on the bottom ground $x_3 = z_0$:

$$A_{13} \partial C/\partial x_3 = - (1-\varepsilon) D \alpha_{13}(K_d C - C^b), \qquad (2.83)$$

where

z_0 is the ground mark,

η is the free surface mark defined by a model of water dynamics (see Section 4.3.4),

ε is the coefficient of ground porosity,

D is the average size of particles,

the constant α_{13} determines the intensity of the adsorption exchange in the system "water - ground deposits",

C^b is the ingredient concentration in ground deposits.

The gravity force acts along the vertical x_3-axis, so the physical processes and properties along the x_3-axis (including the diffusion and advection coefficients v_3, A_3) differ from other two spatial coordinates x_1, x_2.

4.3.2 Equation of Transport of Suspended Particles

Dynamics of the concentration $S(x_1, x_2, x_3, t)$ of the suspended particles in water is described in the diffusion theory framework by the following equation:

$$\partial S/\partial t + \sum_{i=1}^{3} \partial(v_i S)/\partial x_i = \sum_{i=1}^{3} \partial(A_i \partial S/\partial x_i)/\partial x_i - \omega_0 \partial S/\partial x_3, \quad (2.84)$$

where the parameter ω_0 is called *hydraulic size of particles.*

The boundary condition on the free water surface $x_3=\eta$ reflects the condition that the vertical flow is equal to zero if the particle size is ω_0:

(a) at $x_3=\eta$: $$A_3 \partial S/\partial x_3 = (v_3-\omega_0)S, \quad (2.85)$$

The diffusion flow of particles on the bottom is equal to the sedimentation equilibrium flow (see Section 4.3.5) :

(b) at $x_3= z_0$: $$A_3 \partial S/\partial x_3 = -\omega_0 S_*, \quad (2.86)$$

where S_* is the near-ground equilibrium concentration of particles that reflects the transporting capacity of the flow.

4.3.3 Equation of Ingredient Transport on Suspended Particles

The equation of transport of the harmful ingredient on suspended particles with respect to the sought-for ingredient concentration $C^b(x_1,x_2,x_3,t)$ is formulated in the form:

$$\partial(SC^s)/\partial t + \sum_{i=1}^{3}\partial(v_i\, SC^s)/\partial x_i = \sum_{i=1}^{3}\partial(A_i\partial(SC^s)/\partial x_i)/\partial x_i +$$

$$+\ \omega_0\partial(SC^s)/\partial x_3 \ + \alpha_{12}S(K_dC - C^s)), \qquad (2.87)$$

and the corresponding boundary conditions are :

(a) at $x_3=\eta$: $$A_3\partial(SC^s)/\partial x_3 = (v_3-\omega_0)\,(SC^s), \qquad (2.88)$$

(b) at $x_3= z_0$: $$A_3\partial(SC^s)/\partial x_3 + \omega_0(SC^s) = C^s g^s - C^b g^b, \qquad (2.89)$$

where the sedimentation flow g^s and the deposit lift flow g^b are determined in accordance with the direction of overall vertical flow (see Section 4.3.5).

4.3.4 Equations of Surface Water Dynamics

In the general case, the dynamics of surface water is described by the *Reynolds equations for a fluid with free surface*:

$$\partial v_i/\partial t + \sum_{j=1}^{3} v_j\partial v_i/\partial x_j = -(\partial P/\partial t)/\rho + G_i + \sum_{j=1}^{3}\partial\langle v'_i,\, v'_j\rangle/\partial x_i, \qquad (2.90)$$

$$\partial v_i/\partial x_i = 0, \qquad i=1,2,3, \qquad (2.91)$$

where

P is the water pressure,
ρ is the water density,
$\mathbf{G}=(0,0,-g)$ is the gravity force vector,
the 3×3-matrix $\langle v'_i,\, v'_j\rangle$ is a so-called Reynolds turbulent stress tensor.

The *Reynolds turbulent stress tensor* is connected with the *strain velocity tensor* $<(\partial v_i/\partial x_j)(\partial v_j/\partial x_i)>$ by the following formula:

$$<v'_i, v'_j> = -A_{ij}<(\partial v_i/\partial x_j)(\partial v_j/\partial x_i)> - 2\delta_{ij}K/3, \quad i,j=1,2,3, \qquad (2.92)$$

where

the values A_{ij} are called the turbulent change coefficients,
K is the turbulent energy,
δ_{ij} is the *Kronecker delta symbol* ($\delta_{ii}=1$, $\delta_{ij}=0$ at $i=j$).

The equations (2.90)-(2.92) are obviously very general for practical use and require a lot of complex input data. In more practical cases, they are replaced with simpler models of water dynamics (for example, equations (2.96) in Section 4.4).

4.3.5 Equations of Adsorption and Sedimentation

All models mentioned in Section 4.2 use common principles for the description of adsorption change and sedimentation processes. The basic adsorption and sedimentation models are exposed below.

The concentration C^s of the pollutant (a waste product) in the rigid phase is connected with its concentration C in solute by the following equation :

$$dC^s/dt = \alpha(C - \Phi(C^s)), \qquad (2.93)$$

where

α is the coefficient of mass-response in fluid and $\Phi(C^s)$ is *the adsorption isotherm* describing the intensity of the ingredient dissolving in water. Usually, the linear adsorption isotherm is used:

$$\Phi(C^s) = C^s/K_d,$$

where

K_d is the *coefficient of equilibrium distribution* in the system "suspension-water".

Under such description of the adsorption process, the influence of physical-chemical characteristics of water and suspension on the sorption intensity can be defined by the corresponding functions for K_d.

Dynamics of the mass M^b of the dynamic layer of the ground deposits under influence of the suspended particles is described by the equation:

$$dM^b/dt = g^s - g^b, \tag{2.94}$$

where g^s and g^b are the sedimentation and deposit lift flows.

Models of the hydraulic processes of sedimentation and lift of the ground deposits determine the dynamics of the sedimentation flow g^s and the deposit lift flow g^b. The simplest model is of the form:

$$g^s = \begin{cases} k(S - S_*) & \text{at } S > S_*, \\ 0, & \text{at } S < S_*, \end{cases} \qquad g^b = \begin{cases} k(S - S_*) & \text{at } S < S_*, \\ 0, & \text{at } S > S_*, \end{cases}$$

where

S is the concentration of suspended particles (see section 4.3.3 below),
S_* is their equilibrium concentration,
k is a given coefficient.

The set of equations and boundary conditions (2.81)-(2.94) represents a three-dimensional model of water contamination, their different special cases are exposed in next section and (Hritonenko and Yatsenko, 1999).

4.4 Two-Dimensional Horizontal Model and Stationary Flows

Two-dimensional plane (horizontal) model of water pollution propagation is constructed by averaging the equations (2.81)-(2.94) of the three-dimensional model over the water reservoir depth with taking its boundary conditions into account. This model describes the averaged in depth concentration $S(x_1, x_2, t)$ of suspended particles and the concentrations $C(x_1, x_2, t)$, $C^s(x_1, x_2, t)$, $C^b(x_1, x_2, t)$ of the ingredient (harmful substance).

Such planar (horizontal) models are convenient for describing water phenomena in shallow reservoirs (with the scale sufficiently larger than the depth of the reservoir) and for applied problems similar to calculation of flooding plan.

4.4.1 Equation of Ingredient Transport in Dissolved Phase

Integrating the equation (2.81) over the variable x_3 (the flow depth) with taking the boundary conditions (2.82)-(2.83) into account, we obtain the following *two-dimensional equation of the ingredient transport in dissolved phase*:

$$\partial(hC)/\partial t + \sum_{i=1}^{2} \partial(hCv_i) /\partial x_i = \sum_{i,k=1}^{2} \partial(E_{ik} h \partial C/\partial x_i) /\partial x_k -$$

$$- \alpha_{12} S(K_d C - C^s) + \alpha_{13}(K_d C - C^b), \qquad (2.95)$$

where

E_{ik} are the horizontal diffusion coefficients,

$\mathbf{v} = (v_1, v_2)$ is the velocity vector of the water flow averaged over the coordinate x_3 from $x_3=-z_0$ to $x_3=\eta$,

$h= \eta - z_0$ is the flow depth,

z_0 is the ground mark,

η is the free surface mark defined below by the equations (2.98)-(2.99) of water dynamics.

Other notations are the same as in Section 4.3.1. The flow depth $h=\eta-z_0$, the ground mark z_0, and the free surface mark η are obviously dependent on the horizontal spatial coordinates x_1, x_2.

4.4.2 Equation of Suspended Particles Transport

The equation of suspended particles transport obtained by averaging the equation (2.84) over the depth x_3 is of the form:

$$\partial(hS)/\partial t + \sum_{i=1}^{2} \partial(hSv_i)/\partial x_i = \sum_{i,k=1}^{2} \partial(E_{ik}\, h\, \partial S/\partial x_i)/\partial x_k$$

$$- B\omega_0(S_* - S) \qquad (2.96)$$

As compared to the three-dimensional analogue (2.84), a new parameter is the ratio B of the near-ground concentration of particles to the averaged in depth equilibrium concentration S_*, all other notations are the same.

4.4.3 Equation of Ingredient Transport on Suspended Particles

After averaging the equations (2.87)-(2.89), the averaged (in the depth) concentration $C^s(x_1, x_2, t)$ of the ingredient on the suspended particles is described by the following equation:

$$\partial(hSC^s)/\partial t + \sum_{i=1}^{2}\partial(v_i\, hSC^s)/\partial x_i = \sum_{i,k=1}^{3}\partial(E_{ik}h\partial(SC^s)/\partial x_i)/\partial x_k + \\ + \alpha_{12}S(K_dC - C^s)) + C^b g^b - C^s g^s. \quad (2.97)$$

4.4.4 Equations of Water Dynamics

In the planar case, the velocity structure of the water flow can be satisfactory described by the *equations of shallow water theory* :

$$\partial v_i/\partial t + \sum_{j=1}^{2} v_j\partial v_i/\partial x_j + g\partial\eta/\partial x_j = -\lambda v_i|\mathbf{v}| + \lambda_W W_i|\mathbf{W}|,\ i=1,2, \quad (2.98)$$

$$\partial\eta/\partial t + \sum_{j=1}^{2} v_j\partial(hv_i)/\partial x_j = R, \quad (2.99)$$

where

λ is the ground friction (viscosity) coefficient,

λ_W is the coefficient of the wind friction on the free water surface,

$\mathbf{W} = (W_1, W_2)$ is the vector of the wind velocity over water surface,

R is a function of distributed water sources and sinks (influx, sediments, evaporation).

4.4.5 Equation of Ground Deposit Contamination

The change of the thickness $Z(x_1, x_2, t)$ of the ground deposits layer is described by the following *ground strain equation*:

$$(1-\varepsilon)\partial Z/\partial t = g^s - g^b. \tag{2.100}$$

Here the vertical sedimentation flow g^s and the deposit lift flow g^b of the particles are equal to :

$$g^s = \begin{cases} B\varpi_0(S - S_*) & \text{at } S > S_*, \\ 0, & \text{at } S < S_*, \end{cases} \qquad g^b = \begin{cases} B\varpi_0(S - S_*) & \text{at } S < S_*, \\ 0, & \text{at } S > S_*, \end{cases}$$

Then the ingredient concentration $C^b(x_1, x_2, t)$ in the bottom layer of ground deposits is described by the following equation:

$$(1-\varepsilon)\partial(ZC^b)/\partial t = -\alpha_{12}(K_dC - C^b)) - C^b g^b + C^s g^s. \tag{2.101}$$

4.4.6 Analysis of Stationary Flow Problem

As an example of capabilities of the two-dimensional model (2.95)-(2.101) we consider the equations of water dynamics (2.98)-(2.99) under the assumptions that the process is stationary and the wind influence is neglected.

Such problem arises in prediction of spring floods. The purpose of modelling is calculating a spring flood plan. In terms of the specific water expenditures (fluxes) $V_1=v_1h$ and $V_2=v_2h$ the equations (2.98)-(2.99) of water dynamics can be rewritten as :

$$\partial(V_1^2/h)/\partial x_1 + \partial(V_1V_2/h)/\partial x_2 + g\partial\eta/\partial x_1 + \lambda V_1|\mathbf{V}|/h^3 = 0, \tag{2.102}$$

$$\partial(V_1V_2/h)/\partial x_1 + \partial(V_2^2/h)/\partial x_2 + g\partial\eta/\partial x_2 + \lambda V_2|\mathbf{V}|/h^3 = 0, \tag{2.103}$$

$$\partial V_1/\partial x_1 + \partial V_2/\partial x_2 = 0. \tag{2.104}$$

It should be noted that the given depth h is a function $h(x_1, x_2) = \eta(x_1,x_2)-z_0(x_1, x_2)$ of the coordinates x_1, x_2 (as well as the ground mark z_0, and the free surface mark η).

The stationary form of the continuity equation (2.104) allows us to introduce the potential function $\Phi(x_1, x_2)$ called the *stream function* and defined by the formulas:

$$\partial\Phi/\partial x_1 = -V_1, \qquad \partial\Phi/\partial x_2 = V_2, \tag{2.105}$$

The convection terms $\partial(V_1V_2/h)/\partial x_i$ of the equations (2.102) and (2.103) can be omitted when the typical scale of the free surface mark $\chi >> h/\lambda$. Then the cross-differentiation of (2.102) and (2.103) with respect to x_1 and x_2 and the subtraction of the obtained equalities eliminate the terms containing the derivatives $\partial\eta/\partial x_1$, $\partial\eta/\partial x_2$ of the free surface mark η:

$$\partial(V_1|\mathbf{V}|/h^2)/\partial x_1 + \partial(V_2|\mathbf{V}|/h^2)/\partial x_2 = 0, \tag{2.106}$$

Substituting the expressions of V_1 and V_2 from (2.105), we obtain one nonlinear differential equation with respect to the stream function $\Phi(x_1, x_2)$:

$$\partial(F(\mathbf{V})\partial\Phi/\partial x_1)/\partial x_1 + \partial(F(\mathbf{V})\partial\Phi/\partial x_2)/\partial x_2 = 0, \tag{2.107}$$

where the function $F(\mathbf{V}) = -\lambda|\mathbf{V}|/h^3$ depends on $\partial\Phi/\partial x_i$ because of (2.95), and the depth $h(x_1, x_2)$ is a given function of the horizontal spatial coordinates x_1, x_2.

Thus, the system of three partial differential three-dimensional equations (2.98)-(2.99) is reduced to one partial differential two-dimensional equation (2.107). This equation is known as the *quasiharmonic equation*. We can consider the quasiharmonic equation as a non-linear non-uniform generalization of linear elliptic partial differential equations (Laplace equation); see also Appendix.

For further solving the equation (2.107) we need to define a two-dimensional space domain S and the corresponding boundary conditions. Its solution will provide us with the function $\Phi(x_1, x_2)$ whose differentiation with using (2.105) gives the sought-for distribution $V_1=V_1(x_1, x_2)$, $V_2=V_2(x_1, x_2)$ of the water flows for $(x_1, x_2)\in S$.

The common boundary conditions are usually accepted for the stream function $\Phi(x_1, x_2)$ in the calculation of spring floods. Namely, the values Φ_l

and Φ_r are given on rigid boundaries of the domain S, so that $\Phi_r - \Phi_l = Q$ is equal to an estimated spring flood expenditure Q. On fluid boundaries of S, the Dirichlet condition is admitted: $\partial\Phi/\partial n=0$, where n is the normal to the boundary.

Under the given boundary and initial conditions, the quasiharmonic equation (2.107) can be solved with using approximate (analytical or numerical) methods. The analytic theory of the quasiharmonic and similar non-isotropic partial differential equations is a fast developing field of modern mathematics with applications in biology, semiconductor physics, environment protection and other scientific disciplines (Iwaniec and Sbordone, 2001). Numeric methods for such nonlinear equations represent a well-developed specific part of computational mathematics and are beyond the scope of the present book (see remarks in next section).

4.4.7 About Simulation Techniques

There are numerous modifications and improvements of the above general scheme for different applied problems of pollution propagation in environmental systems. Several of them have reached the stage of real information systems and have been used in practical engineering and decision-making activities. Such techniques use stochastic models, differential and difference equations, intelligent algorithms for the identification of parameters, and some other advanced analytical or numerical methods.

An example of a good combination of classical linear and nonlinear differential and difference diffusion models and stochastic identification techniques is the NARMAX model (*Nonlinear Autoregressive Moving Average model with eXogenous inputs*). It describes the related dynamics of rainfall, river levels, and influent flows.

An estimation of a NARMAX model using nonlinear identification techniques is provided in (Tabrizi et al, 2001) for a real-life wastewater treatment plant. The parameters of the model are estimated using a least squares algorithm with intelligent structure detection. The identification makes it possible to predict important relationships between rainfall and river level (inputs) and influent flow (output).

Another good example of applied simulation models is DYMOS (Dynamic Models for Smog Analysis) system that was developed for air pollution analysis (see (Sydow et al, 1998) and the references wherein).

Chapter 3

Variational Models and Structural Stability

This chapter deals with variational techniques applied to the dynamics of mechanical and physical systems and processes. In Chapter 2 we have derived the equations of motion of continuum systems from conservation laws. Variational principles provide an alternative method of obtaining the equations of motion based on the calculus of variations. Also variational techniques are often used in modelling and analysis of new phenomena and constructing approximate methods for solving applied models.

1. VARIATIONAL PRINCIPLES AND MODELS

Variational models are based on *the calculus of variations* - classical mathematical technique developed over two hundred years and applied to many engineering problems (see Section 1.1 in Appendix). There are several variational principles used in classical mechanics (Courant and Hilbert, 1953). One of major and the most universal principles is *Hamilton's variational principle.* It is applicable to a wide class of mechanical systems that have meaningful definitions of kinetic and potential energies (so-called *hamiltonian* systems).

If we can determine expressions for the kinetic energy T and potential energy U for a mechanical or physical system, then the **Hamilton's variational principle** states that *the motion of the system should take place in such a way that the integral of the difference T-U between the kinetic and potential energies is constant.*

The integral produced by the Hamilton's principle (if it exists) usually has a form used in the calculus of variations (for example, see formula (7.1) in Appendix). It is natural because the calculus of variations was developed mostly for mechanical and physical applications. Therefore, the variational techniques may apply to the Hamilton's principle problems. In particular, the necessary conditions of extremum (the *Euler equations* in Appendix) provide a valuable tool for deriving the equations of motion.

Variational techniques are used in applied modelling in different ways. First of all, variational principles provide an alternative method of obtaining the equations of motion. The conservative laws of previous chapter are based on the Newton's laws and require the knowledge of forces acting upon the system. The Hamilton's principle needs to identify kinetic and potential energies only. The variational principles appear to be less universal than the conservative laws; they can't apply to so-called dissipative systems (which are often met in applications). However, this is a classic technology and we illustrate it below in Section 1.1 on classic basic examples of continuum mechanics. Variational models also represent a convenient tool for investigating many spectral problems. We consider some classic and applied variational models for eigenvalue problems in Section 1.2.

Secondly, variational principles are useful in formulating and investigating new models for non-classic applied problems, especially if we are interested in some aggregate structural parameters (i.e., work) rather than in detailed description of distributed characteristics (i.e., displacement) of the process under study. From the viewpoint of pure mathematics, if a variational principle is used as a primary basis for modelling, then it is necessary to consider sufficient conditions of extremum that are more complicated than Euler equations (necessary conditions). However, such conditions may play a little role in the applied problem and may be substituted with appropriate engineering reasoning. We will consider some applied variational models for structural stability problems in Section 2.

Next, the calculus of variations became a basis for various approximate methods (Raleigh-Ritz method, Galerkin method, Kantorovich method, and so on). Such methods are widely used for numeric investigation of applied problems in many areas of engineering and science (Porter and Stirling, 1990). We will touch this aspect of variational methods later.

1.1 Basic Models of Continuum Mechanics

We restrict ourselves with several basic models of continuum mechanics from the classic book (Courant and Hilbert, 1953). Many other examples from particle mechanics, continuum mechanics, optics, physics and others may be found in corresponding textbooks and monographs.

1.1.1 Vibrations of String

Let us consider the process of small transverse vibrations of a homogeneous string fixed at its end points. This is the simplest example of continuum mechanics (see also Section 2 of Chapter 2). The kinetic energy of the string is equal to

$$T = \frac{1}{2}\int_0^L \rho S(\partial u/\partial t)^2\, dx \tag{3.1}$$

where ρ is the mass density and L is the length of the string. The potential energy U can be derived from the work done to increase the length from its value L at rest to its present length $\int_0^L \sqrt{1+(\partial u/\partial x)^2}\,dx$:

$$U = \left(\int_0^L ES\sqrt{1+(\partial u/\partial x)^2}\,dx - ES\right) \approx \frac{1}{2}\int_0^L ES(\partial u/\partial x)^2\,dx. \tag{3.2}$$

In accordance with the Hamilton's principle, the shape of the string is to be such that the integral

$$\int_{t_0}^{t_1}(T-U)dt = \frac{1}{2}\int_{t_0}^{t_1}\int_0^L S\left[\rho(\partial u/\partial t)^2 - E(\partial u/\partial x)^2\right]dxdt \tag{3.3}$$

is stationary over *all* admissible functions $u\,(x,t)$, $x\in[0,L]$, $t\in[t_0,t_1]$. Let us assume that the density ρ and the tension E are constant. Then from the Euler equations (see Appendix, Section 1.1) we obtain the following *wave equation* of the vibrating string:

$$u_{tt} - c^2\, u_{xx} = 0, \qquad c = (E/\rho)^{1/2} \tag{3.4}$$

The equation (3.3) has been earlier obtained in Section 2 of Chapter 2 using the conservation laws.

1.1.2 Transverse Vibrations of Bar

Let us consider the process of small transverse vibrations of a homogeneous bar fixed at its end points. The bar lies in a straight line $y=0$ at rest and subjected to lateral displacements (deflection) $y(x,t)$. The difference between a bar and a string is that the bar is assumed to have a small (comparing with its length) but finite thickness. Therefore, shear forces play an essential role in the bar. The corresponding change in a slope for the neutral plane is $(\partial y/\partial x)^2$ and the elementary deformation (strain) appears to be proportional to $(\partial y/\partial x)^2$. By the Hook's law (see Chapter 2), the stress is proportional to strain and we obtain the following expression for the potential energy U:

$$U \approx \frac{1}{2}\int_0^L EI(\partial^2 u/\partial x^2)^2\,dx. \tag{3.5}$$

where $I=I(x)$ is the moment of inertia of the bar cross-section about an axis perpendicular to the neutral plane of the bar.

Assuming that the kinetic energy T of the vibrating bar is described in the same manner as for the vibrating string in the previous subsection, we obtain from the Hamilton's principle and Euler equations the following *equation of free lateral vibration of the elastic bar*

$$\rho S u_{tt} - \mu u_{xxxx} = 0 \tag{3.6}$$

in the case of the constant density ρ and tension $\mu=EI$ of the bar.

1.1.3 Vibrations of Membrane

Let us consider a plane membrane that covers a region S of the (x,y)-plane at rest and is subjected to small vibrations normal to the (x,y)-plane. The kinetic energy of the membrane is

$$T = \frac{1}{2}\iint_S \rho(\partial u/\partial t)^2\,dx_1 dx_2. \tag{3.7}$$

The potential energy U is equal to the work done by the elastic forces in stretching the membrane from its rest area to its new surface:

$$U = \iint_S E\sqrt{1+(\partial u/\partial x_1)^2+(\partial u/\partial x_2)^2}\,dx_1dx_2 - \iint_S E dx_1dx_2 \approx$$
$$\approx \frac{1}{2}\iint_S E(\partial u/\partial x_1)^2+(\partial u/\partial x_2)^2\,dx_1dx_2. \quad (3.8)$$

In accordance with the Hamilton's principle, the shape of the membrane is such that the integral

$$\int_{t_0}^{t_1}(T-U)dt =$$
$$= \frac{1}{2}\int_{t_0}^{t_1}\iint_S \left[\rho(\partial u/\partial t)^2 - E(\partial u/\partial x_1)^2 - E(\partial u/\partial x_1)^2\right]dx_1dx_2dt$$

is stationary over *all* admissible functions $u(x_1,x_2,t)$, $x\in S$, $t\in[t_0,t_1]$. Let us assume that the density ρ and the tension E are constant. Then from the Euler equations (7.2) in Appendix we obtain the following *equation of the vibrating membrane*:

$$E[(\partial u/\partial x_1)^2+(\partial u/\partial x_1)^2] = \rho(\partial u/\partial t)^2. \quad (3.9)$$

1.1.4 Vibrations of Plate

Let us consider the process of small vibrations of a homogeneous two-dimensional plate. The difference between a plate and a membrane is similar to the difference between a bar and a string in a one-dimensional case. Namely, that the plate is assumed to have a small (if compared with its length) but finite thickness. As usually, the potential energy U of the plate is derived from the work done by elastic forces in returning to the neutral configuration and, in this specific case, is given by an integral of quadratic form in the principal curvatures of the plate. Providing necessary transformations (see (Courant and Hilbert, 1953)) in a manner similar to the vibrating membrane in previous subsection, we obtain from the Hamilton's principle and Euler equations the following *equation of free vibrations of a two-dimensional plate* (with the constant density ρ and tension μ):

$$\rho u_{tt} + \mu \Delta \Delta u = 0. \tag{3.10}$$

For the sake of briefness, we omit the derivation of boundary conditions for a membrane and a plate. Such conditions play an important role in two-dimensional cases and we shall direct the reader to (Courant and Hilbert, 1953) for more details. This classic source also contains many generalizations of the above equations of motions, including influence of external forces and the case of a three-dimensional elastic body.

1.2 Variational Models for Spectral Problems

Spectral problems are concerned with analysis of the so-called *steady-state solutions* to the equations of motion, which are distinct from stationary solutions (equilibrium states). This analysis includes the determination of eigenvalues and eigenfunctions for the equations of motion.

1.2.1 Eigenvalues and Eigenfunctions: Simplest Case

Let us illustrate the concepts of eigenvalues and eigenfunctions for the simple equation of motion of a homogeneous bar built at the left end x=0 and free at the right end x=L. The corresponding boundary problem (2.28)-(2.30) leads to

$$\begin{gathered} u_{tt} - c^2 u_{xx} = 0, \qquad 0 < x < L, \\ u(0, t) = 0, \\ u_x(L, t) = 0. \end{gathered} \tag{3.11}$$

The boundary problem (3.11) is a linear problem with constant coefficients, so we can ask whether it has *harmonic standing wave solutions,* i.e. solutions of the form $u(x,t) = w(x)\exp(i\omega t)$.

Substituting the assumed solution into the problem (3.11), we obtain the following boundary problem for an ordinary differential equation with respect to the sought-for function $w(x)$, $0 < x < L$:

$$w'' + (\omega / c)^2 w = 0, \quad 0 < x < L, \tag{3.12}$$

$$w(0) = dw(L)/dx = 0.$$

It is easy to find the exact solution to this problem :

$$w(x) = C\sin(\omega_n t), \qquad \omega_n = (2n+1)\pi/2L, \quad n=0,1,2, \ldots \tag{3.13}$$

It means that, in general, when $\omega \neq \omega_n$, the problem has only the trivial solution $w=0$ (no motion). However, there are also nontrivial solutions for certain values ω_n of ω. Such values ω_n are known as the *eigenvalues* and the corresponding functions $w_n(x)$ are the *eigenfunctions*. The presence of eigenvalues and eigenfunctions is typical for more general equations.

1.2.2 Raleigh Quotient and Raleigh Method

Let us consider properties of eigenvalues and eigenfunctions for a general linear *conservative* system with the kinetic energy T and the potential energy U. A very useful relationship between the eigenvalue ω and the corresponding eigenfunction $w(x)$ is given by the so-called *Raleigh quotient* (Vierck, 1967; Lin and Siegel, 1974; etc.):

$$\omega^2 = R(w) = T(w)/U(w). \tag{3.14}$$

In the other words, the *Raleigh quotient (the ratio of total potential to kinetic energy for a given mode of vibration) is equal to the square of the eigenvalue.* The equation (3.14) follows from the energy conservation law $T+U$=const. The expressions for the kinetic energy T and the potential energy U depend on the specific equations of motion. For example, by direct substitution of (3.13) into the relation $T+U$=const, one can easily verify that (3.14) is true for the simple motion (3.12)-(3.13).

In the previous example, we calculated the exact eigenvalues and eigenfunctions based on the knowledge of the explicit solution of the problem. Obviously, such a situation is not common in real life situations. A number of techniques have been developed for approximate determination of eigenvalues and eigenfunctions.

The equation (3.14) provides a basis for the *Raleigh method* for approximate finding of the first eigenvalue (the lowest frequency) of a conservative system, which is based on an approximate estimation of the first eigenfunction (the first mode of vibration). This method uses the following variation of the equation (3.14):

$$\omega_R{}^2 = R(u) = T(u)/U(u).$$

where ω_R is called the *Raleigh frequency* and u is an arbitrary admissible function which is assumed be an approximation of the eigenfunction. Practice shows that, when an estimate of an eigenfunction is physically reasonable, the equality (3.14) provides an accurate approximation of the corresponding eigenvalue (see the example in next subsection).

In the Raleigh method, the static equilibrium shape (independent of t) is often used as an estimate u of the first eigenfunction (mode), then the Raleigh frequency gives an approximation of the first eigenvalue (the lowest frequency). This is usually the most important frequency that must be obtained and is often the only information required for many applied systems (Vierck, 1967).

It might be proven by physical reasoning that, in general, the Raleigh frequency ω_R is greater than the correct eigenvalue ω and ω_R is closer to ω for better approximations u. Then the following *minimum principle* may be stated:

$$\omega^2 = R(w) = \min_{u \in W} R(u). \tag{3.15}$$

where ω is the lowest eigenvalue and W is a proper class of all *admissible functions* that includes the eigenfunction w as well. The equality (3.15) shows an example of numerous *extremal properties* of eigenvalues and eigenfunctions that are studied in many advanced texts in mathematical physics, starting with the classic source (Courant and Hilbert, 1953). Below we use the minimum principle (3.15) as a basic equation for finding out a sought-for shape for a bar with certain properties.

1.2.3 Eigenvalues of Bar with Variable Shape

We now consider a bar with the unit thickness and a variable shape in the plan form. The bar is built at the left end $x=L$ and free at the right end $x=L$. Then the bar cross-area $S=S(x)$ is a function of x, $0 \le x \le L$. The longitudinal vibrations of such a bar are governed by the boundary problem (2.28)-(2.30).

Substituting the *harmonic standing wave* solution form $u(x,t) = w(x)\exp(i\omega\ t)$ into this boundary problem, we obtain the following generalization of the boundary problem (3.11) for finding the eigenvalues ω and eigenfunctions $w(x)$:

$$E(S(x)w'(x))' = -\omega^2 \rho Sw, \quad 0 < x < L, \tag{3.16}$$

$$w(0) = w'(L) = 0.$$

which is known as *Sturm-Liouville problem.*

In a general case of the area shape $S(x)$, this problem can not be solved exactly in elementary functions. The *Raleigh quotient* for longitudinal vibrations of a bar in this case is determined as

$$R(u) = \frac{E\int_0^L S(x)(\partial u/\partial x)^2 dx}{\rho\int_0^L S(x)u^2 dx} \tag{3.17}$$

The Raleigh quotient for *transverse vibrations of a bar* (see Section 1.1.2) was derived in (Vierck, 1967) as

$$R_T(u) = \frac{E\int_0^L I(x)(\partial^2 u/\partial x^2)^2 dx}{\rho\int_0^L Su^2 dx}, \tag{3.18}$$

where the moment of inertia of the cross-section at position x about an axis perpendicular to the neutral plane $I=I(x)$ is a function of x, $0\leq x\leq L$, if the bar cross-area $S(x)$ is variable.

The relations (3.17), (3.18) and similar ones may be used in investigation applied engineering problems. First of all, in accordance with the Raleigh method from previous section we can select an approximate function u for the first eigenfunction (mode) w, and obtain an estimate for the first eigenvalue (the lowest frequency). Let us illustrate this point on the following example.

Example. Let us consider a *bar of the wedge form* that is built in at its broad end and free at its pointed end. Then the bar cross-area is $S(x)=C(L-x)/L$. Substituting this formula and the approximation $u=L-x$ for the eigenfunction w into (3.17), we obtain the following approximate value $\omega_R = 2.45(E/\rho)^{1/2}L^{-1}$ for the first eigenvalue (the lowest frequency) for longitudinal vibrations. On the other hand, in this particular case the Sturm-Liouville problem for (3.16) may be resolved in Bessel functions and we obtain the exact first eigenvalue $\omega_R = 2.4(E/\rho)^{1/2}L^{-1}$ (Lin and Segel, 1974). The approximation error is less than 3%.

Other examples of practical interest may be *a stepped beam* whose members being built of different parts or a *beam of telescopic construction.* The Raleigh quotients (3.17) and (3.18) are particularly convenient for investigating such cases. Then the integrals in (3.17) and (3.18) are separated in sums of the integrals over separate portions of the beam.

Note that the first eigenvalue for the uniform homogeneous bar with the constant cross-area is $\omega_R = 1.57(E/\rho)^{1/2}L^{-1}$. We can see that the fundamental frequency (first eigenvalue) increases if a "larger part" of the bar is built in and a "smaller part" is free. This fact relates well to general properties of eigenvalues exposed in (Courant and Hilbert, 1953).

Then the following natural question arises: what shape of the bar provides the lowest (or highest) value of the first eigenvalue? Based on the maximum principle (3.15), we can formulate the exact mathematical statement for this question, which is done in the next subsection. Finding the extremal values of the fundamental frequency may be of a great interest in various engineering problems (for avoiding resonance phenomena).

1.2.4 Extremal Eigenvalues of Bar with Sought-For Shape

Let us consider a bar of the unit thickness, symmetric with respect to the x-axis, and bounded by $S(x)$ and $-S(x)$ in the plane. We assume that the variable shape of the bar $S=S(x)$ is a sought-for function of x, $0 \le x \le L$. The bar is still considered to be a slender (which means that its cross-area is small as compared with its length). Then the model (3.16) for finding the eigenvalues and eigenfunctions of longitudinal vibrations of the bar remains the same. In accordance with the minimum principle (3.15)

$$\omega^2(S) = R(w, S) = \min_{u \in W} R(u,S),$$

where the lowest eigenvalue (fundamental frequency) ω naturally depends on the variable shape S of the bar and the Raleigh quotient $R(u,S)$ is given by the formula (3.17) for longitudinal vibrations.

Now we can formulate the following *optimization problem*:

$$\omega^2_{\max}(S) = \max_{S\in\Omega} \omega^2(S) = \max_{S\in\Omega} \min_{u\in W} R(u,S), \tag{3.19}$$

where Ω is the set of all smooth functions $S(x)$, $0\leq x\leq L$, which satisfy additional boundary conditions at the end points $x=0$ and $x=L$.

Problem (3.19) is a *minimax problem* and can be investigated using various analytical and numeric techniques. The detailed analysis of this and similar problems is behind the scope of this book. We just note that the solution $S(x)$ to the basic problem (3.19) is approximated by an exponential function (*a wedge with exponential boundaries*).

Problem (3.19) determines a bar of a fixed length L with the maximal fundamental frequency. In assumption of the smoothness requirements made above, classical variational techniques can be used for its analysis. This is the simplest example of the extremal problems for eigenvalues. Problem (3.19) may be modified in different directions:

- The condition of a given bar mass (or area) may be used instead of the fixed length.
- Additional constraints may be imposed; for example, the given area of the bar connection to a base (at the built-in end) or a minimum required thickness of the bar at the free end.
- Transverse vibrations may be considered, which means using the Raleigh quotient (3.18) instead of (3.17) in (3.19).
- The condition of the bar thinness may be lifted, then the optimization problem will be formulated for a two-dimensional membrane, where some part of the membrane boundary is fixed (clamped to a base) whereas the rest of the boundary is determined by a sought-for function.

In the last case, the Raleigh quotient is defined by double integrals via the region of the unknown shape S (with some restrictions), i.e. the corresponding optimization problem will be a *problem with sought-for functions in the limits of integration.*

The problems of finding the maximal frequencies play an important role for many engineering applications where some engineering parameters should be extreme for practical reasons. Such problems are related to design of space constructions, military equipment, precise measurement tools, etc. Finding minimal frequencies may be also used in certain engineering applications (it means changing max to min in (3.19)).

Some frequency maximization/minimization problems for simple shapes can be resolved analytically. Such analysis represents a significant part of classical mathematical physics. The following results from (Courant and Hilbert, 1953) give some ideas about the general *extremal properties of eigenvalues*:

- If a system stiffens, the pitch of the fundamental tone and every overtone decreases or remains the same.
- If the inertia of the system stiffens, the pitch of the fundamental tone and every overtone increases or remains the same.
- The fundamental tone of a stretched string of a given uniform tension along which a given mass has been distributed is lowest when the entire mass is concentrated in the middle.
- The membrane of a circled shape has the lowest fundamental tone among all clamped membranes with a given perimeter or area and constant density and elasticity.

2. VARIATIONAL MODELS OF STRUCTURAL STABILITY

Here we consider variational models of structural instability that involve a transition from an equilibrium state to collapse (or another equilibrium state). Such problems consist of finding critical values of some external parameters when a system is loosing its structure or shape. First, a classic mechanical example of such problems, called "buckling rod", from (Courant and Hilbert, 1953) is considered in subsection 2.1. Next, two more specific and complex variational models from capillarity and plasticity areas are analyzed. These examples are adapted from the paper (Strang and Temam, 1982).

It is worth to notice that applied problems of structural stability are often characterized with more aggregated information. Namely, a researcher may be interested in finding out the critical values of some aggregated parameters only (force, energy, work, etc.) rather than in detailed knowledge of distributed characteristics such as stresses and displacements.

2.1 Model of Buckling Rod.

We consider an elastic rod (column) of the length L compressed by a longitudinal force P acting at both ends. The rod is in an equilibrium state. This equilibrium state may be *stable* or *unstable* depending on the value of the force P. In the latter case, a small lateral bending of the rod will cause "buckling", i.e., loosening the present rod structure and switching to a new global state.

While the rod is in the stable equilibrium, it is governed by the variational **principle of minimum potential energy** (Courant and Hilbert, 1953). This principle is a simplification of the Hamilton's variational principle considered in Section 1 for the equilibrium case (without motion). It states that *the potential energy of a system is minimal in the stable equilibrium state.* To formulate a variational model for our problem, we have to consider the total potential energy of longitudinal motion and transverse motion (bending) of the rod. Based on the results of subsections 1.1.1 and 1.1.2, the total potential energy may be described as

$$U \approx \frac{1}{2}\int_0^L EI(u'')^2\,dx - \frac{1}{2}P\int_0^L E(u')^2\,dx. \tag{3.20}$$

where $u(x)$, $0 \le x \le L$, is the lateral (transverse) displacement of the rod from its unstrained position. The first term in (3.20) represents the energy of bending and the second term is the energy of elongation.

Let us assume the simple boundary conditions $u(0)=u(L)=0$ (hinged ends). It is easy to show that for small values of P the minimum of the expression (3.20) is equal to zero and reaches at $u \equiv 0$. On the other hand, for large values of P, the potential energy (3.20) becomes negative. To prove that, it is enough to choose

$$P > \frac{\int_0^L I(u'')^2\,dx}{\int_0^L (u')^2\,dx}$$

for any specific admissible function u. So, for sufficiently large loads P, the state U=0, $u\equiv 0$ is no longer a stable equilibrium.

The buckling force P_{crit} (the largest value of P for which the minimum of the potential energy (3.12) is zero) may be determined by the following variational problem:

$$P_{\text{crit}} = \min_{u \in U} \int_0^L I(u'')^2 \, dx, \tag{3.21}$$

$$\int_0^L (u')^2 \, dx = 1. \tag{3.22}$$

with the boundary conditions $u(0)=u(L)=0$.

Remark. The variational model (3.21)-(3.22) may be also formulated as an eigenvalue problem:

$$(I(x)u''(x))'' + \lambda u''(x) = 0, \quad 0< x<L, \tag{3.23}$$

$$u(0) = u(L) = u''(0) = u''(L) = 0.$$

Then the critical buckling force P_{crit} for the column is equal to the lowest eigenvalue λ in the problem (3.21)-(3.22).

2.2 Model of Anti-Plane Shear Collapse in Plasticity

Let us consider an infinitely long straight pipe, with axis in the z-direction and with the cross-section Ω in the xy-plane (Strang and Temam, 1982) as shown in Figure 3.1. The boundary of the cross-section Ω is Γ. The pipe is subjected to an external load that acts parallel to the z-axis and does not depend on z.

The external uniform force is applied to the surface of the pipe in one direction and to the interior in the other direction (these two forces are balanced so the pipe does not move). Therefore, the only displacement $u(x,y)$ is in the z-direction. This problem is called the *anti-plane shear* because in this case the only stresses in the pipe are the shears $\sigma_1=\sigma_{xz}(x,y)$ and $\sigma_2=\sigma_{yz}(x,y)$.

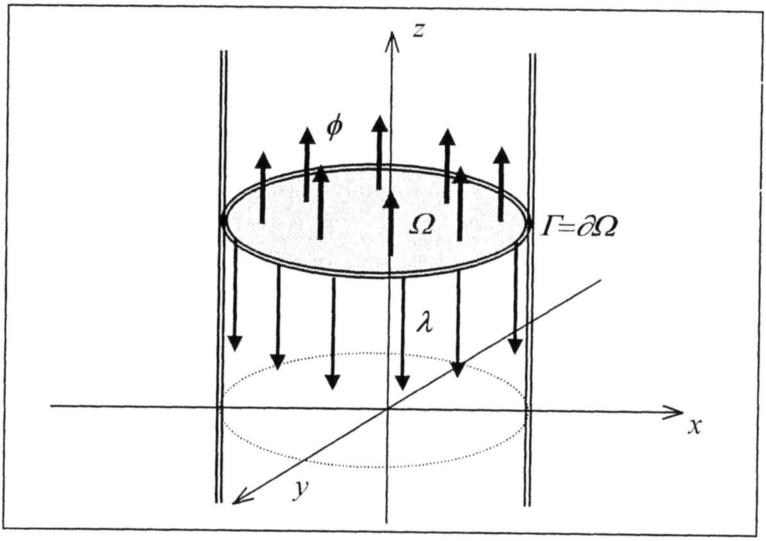

Figure 3.1. Anti-shear problem for a straight pipe (with the unit surface force λ and unit body force ϕ).

A material that the pipe is made of behaves elastically if the total stress is below some critical value σ_{crit} of the stress ('yield stress'), becomes plastic at that point, and cannot accept more stress. As the load increases and the material tries to maintain equilibrium, more and more of the pipe reaches this yield state and a collapse happens at certain *critical value* of the load.

Let us introduce the notation λ for the *surface force* acting along the circumferential length $|\Gamma|$ of the cross-sectional boundary $\Gamma = \partial\Omega$. Then the matching *body force* ϕ acting across the internal cross-sectional area $|\Omega|$ is equal to $\lambda\,|\Gamma|/|\Omega|$ and the corresponding equations of equilibrium are

$$\partial\sigma_1/\partial x + \partial\sigma_1/\partial y = \lambda\,|\Gamma|/|\Omega| \qquad \text{in } \Omega, \tag{3.24}$$

$$\sigma \cdot \mathbf{n} = \lambda \qquad \text{on } \Gamma. \tag{3.25}$$

The equation (3.24) and boundary condition (3.25) are not sufficient to determine σ_1 and σ_2 because there remains a divergence-free stress component, corresponding to zero external forces. To resolve the problem completely, this component must be fixed by an additional relation.

However, at that point we are not interested in the full determination of the stress and displacement distribution.

Another restriction (constitutive relation) comes from the elasticity theory. The hypothesis of perfect plasticity implies that the stresses cannot go outside the 'yield surface' in the stress space. In our case, this condition may written as

$$\sigma_1^2 + \sigma_1^2 \le 1 \quad \text{in } \Omega, \tag{3.26}$$

if the 'yield stress' σ_{crit} is normalized to unity. The inequality holds in the elastic region, and the equality in the plastic region.

The problem of stability analysis is to determine the critical value λ_{crit} of the external force λ at what the conditions (3.24) and (3.25) become incompatible with the constraint (3.26) (it means that the resulting stresses lead to collapse). This value comes from the following convex optimization problem with the sought-for variables λ_{crit}, σ_1 and σ_2:

$$\lambda_{crit} = \max \lambda, \tag{3.27}$$

$$\sigma_1^2 + \sigma_1^2 \le 1 \quad \text{in } \Omega, \tag{3.28}$$

$$\partial\sigma_1/\partial x + \partial\sigma_1/\partial y = \lambda\,|\Gamma|/|\Omega| \quad \text{in } \Omega, \tag{3.29}$$

$$\sigma\cdot\mathbf{n} = \lambda \quad \text{on } \Gamma. \tag{3.30}$$

This optimization problem includes the stress distributions σ_1 and σ_2, which are unknown at this point. Moreover, the stresses cannot be uniquely determined. However, we can find out the optimal λ_{crit} using the duality. As shown in (Strang and Temam, 1982), the dual optimization problem for (3.27)-(3.30) might be written in the form:

$$\lambda_{crit} = \inf_{L(u)=1} \iint_\Omega |\text{grad } u|\, dxdy, \tag{3.31}$$

$$L(u) = \int_\Gamma uds - \frac{|\Gamma|}{|\Omega|} \iint_\Omega udxdy. \tag{3.32}$$

One can see that the optimization problem (3.31)-(3.32) does not include the stresses σ_1 and σ_2. The sought-for variables are λ_{crit} and the corresponding u. Some analysis of the problem (3.31)-(3.32) is provided in the next subsection.

2.3 Model of Capillarity Stability.

Let us consider a straight hollow tube, parallel to the z-axis and with the cross-section Ω in the xy-plane (Strang and Temam, 1982). The tube contains a liquid with the sought-for surface height $u(x, y)$ and the physics of the interface between the liquid and the walls defines the contact angle γ along the cross-sectional boundary $\Gamma=\partial\Omega$. In the absence of gravity, the curvature of the liquid surface is proportional to $\cos\gamma$, and as the curvature increases some part of the surface becomes climbing the walls as shown in Figure 3.2. At some critical value of $\cos\gamma$ the surface breaks down and a certain part of the surface climbs to infinity (which means the collapse). The angle γ is governed by the surface tension and is considered to be a sought-for parameter in the structural stability problem.

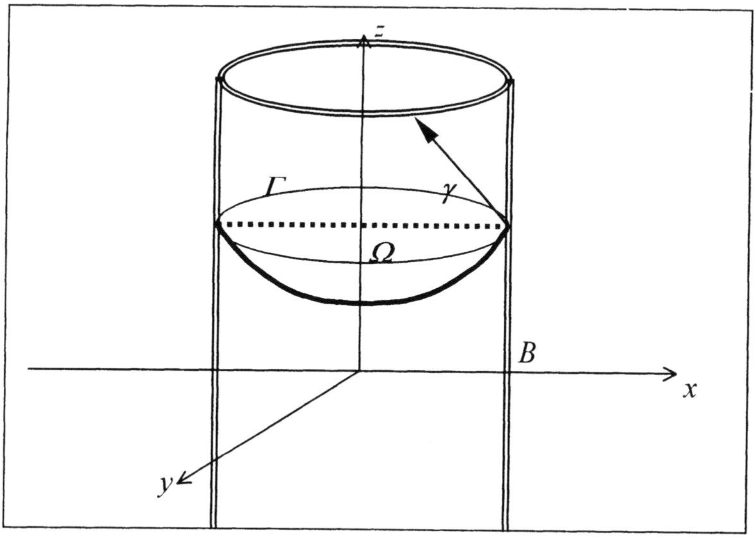

Figure 3.2. A liquid in the thin hollow tube B climbs the walls because of capillarity effect (the bold dashed line shows the flat liquid surface in the case $\gamma=\pi/2$ – no capillarity).

The angle γ is measured between the normal to the surface and the normal to the wall, so that $\gamma = \pi/2$ corresponds to the case of flat surface without capillarity effects. We assume that $\gamma \leq \pi/2$. The geometry provides the following equality for $\cos\gamma$ on the boundary Γ:

$$\cos\gamma = \frac{(u_x, u_y, 1)\cdot(n_x, n_y, 0)}{(1 + u_x^2 + u_y^2)^{1/2}}.$$

Let us denote the denominator $(1 + |\nabla u|^2)^{1/2}$ as W, then the equality becomes

$$\frac{\mathrm{grad}u \cdot \mathbf{n}}{W} = \cos\gamma, \qquad (x,y) \in \Gamma. \tag{3.33}$$

In the interior, the surface of liquid obeys the *Young-Laplace equation* :

$$\mathrm{div}\left(\frac{\mathrm{grad}u}{W}\right) = 2H, \qquad (x,y) \in \Omega - \Gamma. \tag{3.34}$$

Equality (3.34) means that the mean curvature H of the liquid surface is constant.

We recognize in (3.33)-(3.34) the Euler equations and natural boundary condition (see Appendix) for the following variational problem:

$$\iint_\Omega W\,dxdy - \cos\gamma \int_\Gamma uds \longrightarrow \min_{0\leq\gamma\leq\pi/2}, \tag{3.35}$$

$$\iint_\Omega udxdy = C. \tag{3.36}$$

The constant mean curvature H is determined from γ by the divergence theorem:

$$\iint_\Omega 2H\,dxdy = \int_\Gamma \cos\gamma ds, \quad \text{or} \quad 2H = \frac{|\Gamma|}{|\Omega|}\cos\gamma. \tag{3.37}$$

We will use the constant H as the Lagrange multiplier for the constraint equality (3.36). After applying the Lagrange multiplier method and using (3.37), the variational problem (3.35)-(3.36) becomes:

$$\iint_{\Omega} W\, dxdy - L(u)\cos\gamma \longrightarrow \min_{0\le\gamma\le\pi/2}, \tag{3.38}$$

where

$$L(u) = \int_{\Gamma} uds - \frac{|\Gamma|}{|\Omega|}\iint_{\Omega} udxdy. \tag{3.39}$$

The problem is close to the well-known Plateau's minimal surface problem. In the trivial case $\gamma = \pi/2$, $\cos\gamma = 0$ (no capillarity effects), the minimum to the problem (3.38)-(3.39) is delivered by the flat surface $u\equiv$const and $L(u)$ vanishes for this u.

The goal of our *structural analysis* is to decide whether the minimization problem (3.38)-(3.39) is bounded or unbounded, i.e. whether the minimum is $-\infty$. This condition is used as the test for collapse. It is reasonable to expect that a critical value $\cos\gamma_{crit}$ of the sought-for control parameter $\cos\gamma$ exists such that the minimum is finite and a solution of the initial boundary problem (3.33)-(3.34) may be constructed for $\cos\gamma<\cos\gamma_{crit}$. For $\cos\gamma>\cos\gamma_{crit}$, the term $L(u)\cos\gamma$ brings the minimum to $-\infty$ and the equations (3.33)-(3.34) have no solution.

Since we test only the 'yes-no' question, this allows us to simplify the integrand with replacing the function W by the gradient $|\nabla u|$. More exactly, we can note that

$$|\nabla u| < (1 + |\nabla u|^2)^{1/2} = W \le 1 + |\nabla u|,$$

i.e., the values of W and $|\nabla u|$ differ by less than 1. Therefore, the integrals of W and $|\nabla u|$ over Ω will differ by less than a fixed constant, namely the area of Ω:

$$\iint_{\Omega} |\nabla u| dxdy < \iint_{\Omega} W dxdy \le \iint_{\Omega} (1 + |\nabla u|) dxdy = |\Omega| + \iint_{\Omega} |\nabla u| dxdy.$$

It means that the minimum remains finite or infinite when we replace the minimization problem (3.38)-(3.39) with the following problem:

$$\iint_{\Omega} |\nabla u|\, dxdy - L(u)\cos\gamma \longrightarrow \min_{0\le\gamma\le\pi/2}, \tag{3.40}$$

The last problem is homogeneous: if u is multiplied by $\alpha>0$, then so is the value of the functional. Therefore, if the functional is negative for any u, then its minimum is $-\infty$. In other words, the critical value of collapse is determined by the following minimization problem:

$$\cos\gamma_{\text{crit}} = \inf_{L(u)=1} \iint_{\Omega} |\nabla u|\, dxdy, \tag{3.41}$$

$$L(u) = \int_{\Gamma} u ds - \frac{|\Gamma|}{|\Omega|} \iint_{\Omega} u dxdy. \tag{3.42}$$

For $\cos\gamma$ below $\cos\gamma_{\text{crit}}$, the minimum in the optimization problem (3.40) is positive and finite. For $\cos\gamma>\cos\gamma_{\text{crit}}$, the functional in (3.40) is negative for some u and then the minimum (3.40) is $-\infty$.

One can notice that the final version (3.41)-(3.42) of the mathematical statement of the above capillarity problem is the same that the dual problem (3.31)-(3.32) for the plasticity problem considered in Section 2.2.

In conclusion, we would like to mention one result about the solutions of variational problems (3.31)-(3.32) and (3.41)-(3.42), which greatly simplifies the analysis of the corresponding applied engineering problems.

It is convenient to consider the variational problem (3.41)-(3.42) in the space $\boldsymbol{BV}_{\Omega}$ of the functions of bounded variation on Ω (see Section 1.1 in Appendix). Roughly, this is the largest class of functions u for which $\iint_{\Omega} |\nabla u|\, dxdy < \infty$. This space contains, in particular, *characteristic functions* $\chi_E(x,y)$ for all reasonable sets E in Ω. We remember that the characteristic function $\chi_E(x,y)=1$ for $(x,y)\in E$ and $\chi_E(x,y)=0$ for $(x,y)\notin E$.

As shown in (Strang and Temam, 1982), the solution of the problem (3.41)-(3.42) is delivered by the characteristic functions. Namely, there is always a minimizing sequence of characteristic functions, and if the minimum is attained, it is attained at a characteristic function. For example, if Ω is the unit square, then the minimizing solution is χ_E where E is a small triangle in the corner of the square when the perimeter $|E|$ of the triangle tends to 0.

The minimizing characteristic function χ_E has a *clear physical interpretation* for both capillarity and plasticity problems. For capillarity, the boundary ∂E of the area E will separate the part of the surface that remains finite from the part of E that climbs to infinity. For plasticity, it represents the "hinge line" across which the pipe will shear (when the pipe breaks, the part with the cross-section E moves away).

Chapter 4

Integral Models Of Physical Systems

Integral equations represent a more general tool for solving applied engineering problems as compared with differential equations. An integral equation "represents the entire physics of the problem in a very compact form and, in many instances, a more convenient form than the more conventional differential equation" (Morse and Feshbach, 1953, p.896). Generally speaking, integral equations (IEs) can describe global situations that cannot be modeled by differential equations. On the other hand, all models based on DEs may be converted to IEs. The corresponding applications include many problems of viscoelasticity (Renardy et al, 1987) and creep theory (Arutjunian and Kolmanovskii, 1983), superfluidity (Miller, 1973) and aeroelasticity (Belotserkovskii et al, 1980), coagulation and meteorology (Galkin and Dubovskii, 1982), electromagnetism (Bloom, 1981), radiation transfer (Cergignani, 1975; Colton and Kress, 1983), radiophysics (Ramm, 1980), electronic lithography, astronomy, and so on.

We have already met several examples of IEs in previous chapters. This chapter concentrates on integral models and their features.

The first section is devoted to the construction of integral models based on the conversion of differential models as well as on a direct integral description of a process under study. Two powerful IE methods are considered there: Green's functions and method of boundary integral equations. Following sections deal with some applied integral models in engineering acoustics and mining engineering. In Section 2, the method of boundary integral equations is applied to a problem of environmental noise prediction due to a traffic road (cars, train) over an inhomogeneous ground grassy area. An integral model of a mine elevator considered in Section 3 describes the dynamics of a mine rope as a one–dimensional continuum of variable length.

1. CONSTRUCTION OF INTEGRAL MODELS

There are two major techniques for obtaining applied integral models:

- Reduction of DEs to IEs that gives some advantages in many practical cases. The corresponding methods have been developed during centuries and are discussed in Section 1.1.
- Direct construction of integral models based on certain integral conservation and constitutive relations. There are many situations that cannot be described in terms of DEs because of their global physical character but they can be of being modeled via IEs (Section 1.2).

1.1 Converting Differential Models to Integral Models

Integral models are more general but differential models are simpler and more common in analytical and numerical analysis. A general rule is: if a differential model can *effectively* describe a process with a required accuracy, then there is no need in constructing and using integral models.

However, in many applied situations, differential models meet significant difficulties in their investigation, which are connected with complicated boundary geometry or complicated boundary conditions, instability of corresponding numerical algorithms, etc. As mentioned in the classic source (Morse and Feshbach, 1953), "the scattering of waves from an object which is not quite a sphere, the propagation of waves in a duct which is not everywhere uniform ... cannot usually be solved exactly... For such problems the use of Green's functions, of integral equations, and of variational principles offers methods of utility and power."

The advantages of the integral models include:

- better flexibility for accounting new features and phenomena;
- possible reduction in the dimension of a problem (for example, a two-dimensional DE is often reduced to a one-dimensional IE), see BIE method of Section 1.4;
- established techniques for proving existence and uniqueness of solutions, finding eigenvalues and eigenvectors, etc.;
- uniform structure of models and corresponding approximate methods;

- better stability of the numerical methods; and so on.

Transforming DE to an equivalent IE is a classic topic and has been thoroughly explained in many monographs and textbooks (Burton, 1983; Jerri, 1985; Corduneanu, 1991; Gripenberg et al, 1990; and so on). We will restrict ourselves with several basic examples for the initial value and boundary value problems for ordinary and partial differential equations (ODE and PDE). The major difference between these cases lies in the following:

- an initial value DE problem is converted to an integral equation with variable limits of integration (to Volterra IE in the case of a linear DE);
- a boundary value DE problem is converted to an integral equation with constant limits of integration (to Fredholm IE in the linear case).

Volterra and Fredholm IEs differ significantly in their analytical theory and methods of investigations. One can find more detailed comparison of DEs and their integral analogues in Appendix (Chapter 7). Converting a DE to an equivalent IE is also used for more specific applied problems in Sections 2 and 3.

1.1.1 Initial Value Problems

Ordinary Differential Equations. It is well known that the linear ordinary differential equation (ODE) of n-th order:

$$\frac{d^n x}{dt^n} + a_1(x)\frac{d^{n-1}x}{dt^{n-1}} + \dots + a_{n-1}(x)\frac{d^{n-1}x}{dt^{n-1}} + a_n(x)x(t) = F(t)$$

with the initial conditions

$$x(0) = c_0, \quad \frac{dx(0)}{dt} = c_1, \ \dots, \quad \frac{d^{n-1}x(0)}{dt^{n-1}} = c_{n-1},$$

is equivalent to the following linear *Volterra integral equations of the second kind* with respect to the function u:

$$u(t) = \int_0^t K(t,\tau)u(\tau)d\tau + f(t)$$

where

$$K(t,\tau) = \sum_{k=1}^{n} a_k(t)\frac{(t-\tau)^{k-1}}{(k-1)!},$$

$$f(t) = F(t) - \sum_{k=1}^{n} a_k(t)\sum_{l=1}^{k} \frac{c_{n-l}t^{k-l}}{(k-l)!}.$$

Automatic control theory (Barnett, 1975; Brogan, 1974; Eykhoff, 1974; Lee and Markus, 1967; Ljung, 1987; Soderstrom and Stoica, 1974; and others) provides a good reason for converting differential equations to integral ones because this engineering branch regularly uses both integral and differential models of *dynamic systems* (DS).

An *open-loop input-output model* of a linear dynamic system is described by the system of ODEs of the first order:

$$dx/dt = A(t)x + G(t)u, \tag{4.1}$$

where the sought-for *output signal* $x(t)=(x_1, x_2, ..., x_n)$ is an n-vector-valued function, the given *input signal* $u(t)=(u_1, u_2, ..., u_m)$ is an m-vector-valued function, A and G are matrix-valued functions of corresponding dimensions.

The analytical solution of the initial value problem $x(t_0)=x_0$ for ODE (4.1) is of the following form:

$$x(t) = \Phi(t,t_0)x_0 + \int_{t_0}^{t} \Phi(t,\tau)G(\tau)u(\tau)d\tau, \tag{4.2}$$

where a so-called *transition matrix* $\Phi(t,\tau)$ is defined by the differential equation: $\partial\Phi(t,\tau)/\partial t = A(t)\Phi(t,\tau)$, $\Phi(t,t)=E$, E is the identity matrix.

Equality (4.2) defines an explicit connection between the input and output and represents an *integral model* of the open-loop linear DS, which is equivalent to ODE (4.1). We can rewrite model (4.2) as:

$$x(t) = f(t) + \int_{0}^{t} K(t,\tau)u(\tau)d\tau. \tag{4.3}$$

If the DS is *stationary* (its parameters do not depend on time t), then $A(t)\equiv A$, $G(t)\equiv G$ in differential model (4.1) and $K(t,\tau)=K(t-\tau)$ in integral model (4.3) (the output $x(t)$ depends only upon the time $t-\tau$).

Another commonly used explicit integral model of linear DS is described by the of the following form:

$$x(t) = \int_{-\infty}^{t} K(t,\tau)u(\tau)d\tau, \tag{4.4}$$

where $K(t,\tau)$ is called *the unit impulse response (UIR)* of DS. If the function $K(t,\tau)\neq 0$ for all $t-\tau>0$, then DS is referred to as *the DS with infinite memory*. In many practical cases, DS output $x(t)$ depends on the input signals $u(\tau)$ whose time instants τ are no farther from t than a critical distance $T>0$, i.e. $K(t,\tau)\equiv 0$ for $t-\tau>T$. The value T is called *the after-effect duration* or *the DS memory*. In the last case we have *the DS with finite memory* modeled as

$$x(t) = \int_{t-T}^{t} K(t,\tau)u(\tau)d\tau, \tag{4.4'}$$

Let us suppose that the structure of linear DS (4.1) is defined by the additional linear relation between the input u and output x:

$$u(t) = Y(t)x(t)+u_0(t). \tag{4. 5}$$

The coupling types and intensities are determined by the $n\times m$-matrix $Y(t) = \{y_{ij}(t)\}$. If the DS has no active elements (amplifiers), then $|y_{ij}(t)|\leq 1$ for all i,j. Equality (4.5) represents a *feedback control law* for the DS. If $y_{ij}(t)\neq 0$, then there is the *feedback* in the output channel i (positive at $y_{ij}(t)>0$ and negative at $y_{ij}(t)<0$). Different types of structural connections among the system elements can be described with the appropriate choice of $Y(t)$, in particular, parallel and series connections.

Substituting (4.5) into the explicit IM (4.3), we obtain the following *implicit IM* of the linear DS:

$$x(t) = \int_{0}^{t} K(t,\tau)Y(\tau)x(\tau)d\tau + \hat{f}(t), \tag{4.6}$$

where $\hat{f}(t) = \int_0^t K(t,\tau)u_0(\tau)d\tau + f(t)$ is the given DS input. Model (4.6) represents the system of *Volterra integral equations of the second kind* with respect to the output $x(t)$. This model is equivalent to *the closed-loop linear differential model* obtained by substituting (4.5) into ODE (4.1).

If, instead of (4.5), we assume the *non-linear feedback law* of the form:

$$u(t) = F(t, x(t)), \tag{4.7}$$

then we obtain the following *nonlinear integral equation with the variable limit of integration* (*Volterra-Hammerstein equation*) in the variable $x(t)$:

$$x(t) = \int_0^t K(t,\tau)F(\tau, x(\tau))d\tau + \tilde{f}(t). \tag{4.8}$$

Nonlinear integral model (4.8) is equivalent to *the closed-loop nonlinear differential model*:

$$dx/dt = A(t)x + G(t)F(t, x(t)), \tag{4.9}$$

obtained by the substitution of feedback relation (4.7) into the ODE system (4.1).

A nonlinear DS can be described by a more general nonlinear ODE:

$$dx/dt = \Phi(t, x, u), \tag{4.10}$$

where F is an n-vector-valued function of $n+m+1$ variables t, x_1, x_2, ..., x_n, u_1, u_2, ..., u_m. The initial value problem $x(t_0)=x_0$ for ODE (4.10) is known to be equivalent to the nonlinear integral equation:

$$x(t) = \int_0^t \Phi(\tau, x(\tau))d\tau + x_0. \tag{4.11}$$

One can see from the comparison of the nonlinear models (4.9) and (4.10) that the transformation of DE into IE can be done in different ways (of course, the obtained IEs will be equivalent). For the briefness sake, we omit the questions of solution existence. Necessary assumptions of smoothness should be done to validate the used procedures.

Partial Differential Equations. A statement of an initial value problem has sense for some types of dynamic partial differential equations as well; for example, for the wave equation or diffusion equation (see Chapters 2 and 3). The *type of integral equations obtained* depends on the physical character of a process.

To illustrate possible cases, let us we consider a "pure" initial value problem for the following *two-dimensional quasilinear hyperbolic equation*:

$$u_{xt} = F\,(x,t,u), \tag{4.12}$$

in the first quadrant $x \geq 0$, $t \geq 0$, with the following initial conditions:

$$u(x,0) = u_1(x), \qquad u(0,\,t) = u_2(t). \tag{4.13}$$

Using direct integration with respect to variables x and t, we obtain the following nonlinear *Volterra integral equation of the second kind*:

$$u(x,t) = u_0(x) + u_2(t) - u_2(0) + \int_0^x \int_0^t F(\xi,\tau,u(\xi,\tau))d\xi d\tau, \tag{4.14}$$

which is equivalent to problem (4.12)-(4.13) under corresponding requirements of smoothness. The described process is a pure wave propagation with no reflection or distribution phenomena.

Let us now we consider an initial value problem for the following *two-dimensional quasilinear diffusion equation*:

$$u_t = u_{xx} + f(x,t,u), \tag{4.15}$$

in the half-space $t \geq 0$, with the initial condition:

$$u(x,0) = u_0(x), \qquad x \in (-\infty,\infty). \tag{4.16}$$

Following (Corduneanu, 1991), this problem is converged to the following two-dimensional nonlinear integral equation:

$$u(x,t) = \int_{-\infty}^{\infty} G(t, x-\xi)u_0(\xi))d\xi +$$
$$+ \int_0^t \int_{-\infty}^{\infty} G(t-\tau, x-\xi)F(\xi,, \tau, u(\xi,\tau))d\xi d\tau. \tag{4.17}$$

where $G(t,x) = [\pi\delta(t-s)]^{-1/2}\exp[-x^2/4(t-s)]$ is the Green's function for the linear PDE $u_t = u_{xx}+\delta(x)$ (see also Section 1.2 below). The process under study includes a diffusion phenomenon together with the distribution along the space coordinate x. Correspondingly, equation (4.17) is *of the Volterra type* in the time variable *t and of Fredholm type* in the space variable x.

Applied dynamic spatial models described by PDEs are usually considered in a restricted space region and accompanied by boundary conditions. The corresponding processes may include wave propagation, distribution, refraction, reflection, space diffusion, and other phenomena. Being transformed to IEs, such problems usually generate mixed multidimensional integral equations similar to (4.16) that are of the Volterra type in the time variable and of Fredholm type in space variables.

1.1.2 Boundary Value Problems for Ordinary Differential Equations: Green's Function

Transformations of boundary value and initial value problems into integral equations are different because the theory of boundary value problems is much more complex as compared with the initial value problems. Correspondingly, we need to consider more specific cases. The most studied ODE boundary value problem is a two-point boundary problem for a linear ODE of the second order with variable coefficients, which is called *the Sturm-Liouville problem* (see also Section 4.2.3 of Chapter 3). An exhaustive analysis of transformation of the Sturm-Liouville boundary value problem (including its singular case) to integral equations is provided in (Porter and Stirling, 1990).

A classic technique for transforming boundary value problems to integral equations is based on the *Green's function* (or *influence function*). The Green's function is usually introduced as a solution of a *homogeneous* boundary value problem for the corresponding *homogeneous* DE. It represents the kernel of an equivalent integral equation. There are various techniques for construction of the Green's functions; some of them are illustrated for the PDE case in next subsection. Here we restrict ourselves with the following simple but nonlinear two-point boundary problem for the ODE of the second order (Porter and Stirling, 1990):

$$y''(x) = F(x, y(x)), \qquad 0< x<1, \tag{4.18}$$

$$y(0) = y_0, \qquad y(1) = y_4.$$

After double integration of equation (4.18) and determination of the unknown constants of integration from the boundary conditions, we obtain the following *nonlinear integral equation with constant limits of integration* (the *Hammerstein equation*) in the variable $x(t)$:

$$y(x) = -\int_0^1 K(x,t)F(t, y(t))dt + f(x), \tag{4.19}$$

where

$$K(x,t) = \begin{cases} t(1\text{-}x), & t \le x, \\ x(1\text{-}t), & x \le t, \end{cases} \tag{4.20}$$

$$f(x) = (y_1 - y_0)x + y_1.$$

We can reverse the process and obtain the boundary problem (4.18) from the integral equation (4.20). So, these problems are equivalent.

The kernel $K(x, t)$ represents the Green's function for the linear *homogeneous case* $y''(x) = \delta(0)$, $0< x<1$, $y(0)=y(1)=0$, of the nonlinear boundary problem (4.18). Here the *Dirac delta-function* $\delta(.)$ represents a unit singular point force at $x=0$.

It is important to notice that the Green's function depends on the type of boundary conditions rather than on their specific values. For example, if we change the nonlinearity $F(x,y(x))$ to the linear function $F(x,y)=y$, then the integral equation (4.18) becomes the classical linear *Fredholm equation of the second kind,* with the kernel of the same form (4.20).

However, if we replace the boundary conditions in the nonlinear two-point problem (4.18) with $y(0)=y'(0)=0$, then the Green's function becomes:

$$K(x,t) = \min(x, t) \tag{4.21}$$

The Green's function (3.15) will be used in an applied model of Section 2.

1.1.3 Boundary Value Problems for Partial Differential Equations: Boundary Integral Equation Method

In the case of PDEs, boundary conditions play a more important role in investigation of the problem as compared with ODEs. Correspondingly, the elementary approach used for converting ODE boundary problems is not sufficient and more sophisticated techniques have been developed for PDE boundary problems. These techniques are known as the *method of boundary integral equations* (*BIE method*). It uses and develops the classic concept of the Green's function (see previous section) that is introduced as the kernel of an equivalent integral equation.

The essence of the BIE method for solving boundary problems (Maz'ya, 1989) lies in a special representation of solutions of partial differential equations as integrals over the domain involving one or more unknown functions on the boundary of the domain. The boundary conditions are used to determine these functions. Substituting the obtained representation into the boundary operators converts the boundary conditions into the integral (or integro-differential) equations on the boundary. By solving the latter equations, one can obtain a solution of the original boundary problem over the whole domain. In doing so, the dimension of the obtained boundary integral (or integro-differential) equation is one less than of the original PDE boundary problem.

In this section, we illustrate the BIE method for the case of the *Helmholtz equation:*

$$\Delta u + k^2 u = f(\mathbf{x}), \quad \mathbf{x} \in \mathrm{R}^n, \quad n=1,2,3, \tag{4.22}$$

which is the basic equation of engineering acoustics. It describes the sound pressure field due to harmonic excitation. The obtained results will be used in an applied acoustic model of Section 3.

Although the equation (4.22) is linear and may be investigated by various analytical methods, such methods are usually applicable for simple geometric regions only. We would like to emphasize that the BIE method provides the most general representation of the acoustic field for an arbitrary geometric configuration (Filippi et al, 1999).

Basically, the BIE method includes the following steps:

- determining the Green's function (fundamental solution);
- defining a presentation for the sought-for solution inside the region using the Green's function and fictitious sources on the boundary;
- obtaining a BIE on the boundary for the unknown fictitious sources;
- calculation of the solution in the whole region via the obtained presentation.

We will consider the following boundary value problem for the equation (4.22) in a bounded domain $\Omega \subset \mathrm{R}^n$ (n=1, 2 or 3) with a closed boundary σ:

$$(\Delta + k^2)u(\mathbf{x}) = f(\mathbf{x}), \qquad \mathbf{x}\in\Omega, \tag{4.23}$$

$$\alpha\partial_n u(\mathbf{x}) + \beta u(\mathbf{x}) = 0, \qquad \mathbf{x}\in\sigma, \tag{4.24}$$

The boundary condition (4.24) is known as the *Robin boundary condition* and is widely used in engineering acoustics. The well-known Neumann and Dirichlet boundary conditions are obtained from (4.24) at β=0 and α=0 correspondingly. The Robin condition (4.24) describes a wide range of practical situations, from a perfectly rigid solid boundary that cuts the normal velocity component (the Neumann condition) to a liquid boundary with zero acoustic pressure on it like the sea surface (the Dirichlet boundary condition). An intermediate case $\beta\neq 1$, $\alpha\neq 1$ corresponds to boundaries that partially adsorb the acoustic energy and partially transmit it.

We will omit the existence and uniqueness of a solution to problem (4.23)-(4.24) and refer an interested reader to (Courant and Hilbert, 1953).

First of all, we need to determine the Green's function. Various definitions of the Green's function depending on the imposed conditions are possible for equation (4.22). We will use the simplest *free space Green's function for the Helmholtz equation* defined as the solution to the following equation

$$(\Delta + k^2)u(\mathbf{x}) = \delta_{\mathrm{y}}(\mathbf{x}), \tag{4.25}$$

that satisfies the so-called *Sommerfeld conditions*

$$u \sim O(r^{(1-n)/2}), \qquad \partial_r u - iku \sim o(r^{(1-n)/2}), \tag{4.26}$$

when the distance $r(\mathbf{x},\mathbf{y})$ between points $\mathbf{x}$ and $\mathbf{y}$ moves toward ∞. Here the *Dirac delta-function* $\delta_{\mathrm{y}}(\mathbf{x})$ represents the point isotropic source of the acoustic wave located at point $\mathbf{x} \in \mathsf{R}^n$. It may be verified by a direct substitution that the Green's function has the form of

$$G(\mathbf{x},\mathbf{y}) = -\exp[ikr(\mathbf{x},\mathbf{y})]/4\pi r(\mathbf{x},\mathbf{y}), \quad \mathbf{x} \in \mathsf{R}^3, \tag{4.27}$$

$$G(\mathbf{x},\mathbf{y}) = iH_0^{(1)}[kr(\mathbf{x},\mathbf{y})]/2ik, \quad \mathbf{x} \in \mathsf{R}^2, \tag{4.27a}$$

$$G(\mathbf{x},\mathbf{y}) = \exp[ikr(\mathbf{x},\mathbf{y})]/2ik, \quad \mathbf{x} \in \mathsf{R}^1, \tag{4.27b}$$

where $H_0^{(1)}(z)$ is the *Hankel function* of order zero of the first kind. The function $G(\mathbf{x},\mathbf{y})$ describes the radiation of a small isotropic source in free space. Then the solution of the Helmgoltz equation (4.22) *in free space* may be described by the following integral

$$u_f(\mathbf{x}) = \int_{\mathbf{R}^n} G(\mathbf{x},\mathbf{y}) f(\mathbf{y}) d\Omega(\mathbf{y}). \tag{4.28}$$

Formula (4.28) represents a so-called *incident pressure* (the pressure that the sources $f(\mathbf{x})$ generate in free space).

Let us represent the sought–for sound pressure $u(\mathbf{x})$ in the boundary problem as the sum of two pressures:

a) the incident pressure $u_f(\mathbf{x})$ (4.28),
b) the *reflected* (or *diffracted*) pressure $u_r(\mathbf{x})$ due to fictitious sources on the boundary σ (the sources should be chosen to satisfy the boundary condition (4.24) for the total pressure $u(\mathbf{x})$).

Various representations are possible for the reflected sound pressure $u_r(\mathbf{x})$ in the propagation domain Ω. A classical representation for the total pressure $u(\mathbf{x})$ (the *Green representation*) in R^2 and R^3 is given by the formula:

$$u(\mathbf{x}) = u_f(\mathbf{x}) - \int_\sigma \left[\partial_{n(\mathbf{y})} u_g(\mathbf{y}) G(\mathbf{x}, \mathbf{y}) - u_g(\mathbf{y}) \partial_{n(\mathbf{y})} G(\mathbf{x}, \mathbf{y})\right] d\sigma(\mathbf{y}), \quad (4.29)$$

where $u_g(\mathbf{x})$, $\partial_n u_g(\mathbf{x})$ are unknown densities of additional sources on the boundary σ. These two functions are related through the boundary condition (4.24). It may be shown that the constructed pressure (4.29) is equal to the solution $u(\mathbf{x})$ of problem (4.23)-(4.24) for $\mathbf{x}\in\Omega$ and is zero outside the domain Ω. It means that the solution $u(\mathbf{x})$ will be discontinuous through the boundary σ.

Assuming $\alpha\neq 0$ and using the boundary condition (4.24), we can express the normal derivative of the pressure $\partial_n u_g(\mathbf{x})$ in terms of the pressure $u(\mathbf{x})$:

$$\partial_n u(\mathbf{x}) = -\beta u(\mathbf{x})/\alpha \quad \text{for } \mathbf{x}\in\sigma.$$

Then the Green representation may be written as

$$u(\mathbf{x}) = u_f(\mathbf{x}) + \int_\sigma u_g(\mathbf{y}) \left[\beta G(\mathbf{x}, \mathbf{y})/\alpha + \partial_{n(\mathbf{y})} G(\mathbf{x}, \mathbf{y})\right] d\sigma(\mathbf{y}), \quad (4.30)$$

where $\mathbf{x}\in\Omega$.

Finally, the boundary integral equation for $u_g(\mathbf{x})$ is obtained from the integral representation (4.30) by letting a point $\mathbf{x}\in\Omega$ in the propagation domain Ω tend to a point $\mathbf{x}\in\sigma$ on the boundary σ. Taking into account the discontinuity of the solution $u(\mathbf{x})$ through the boundary σ, we obtain the following integral equation for $u_g(\mathbf{x})$, $\mathbf{x}\in\sigma$:

$$u_g(\mathbf{x})/2 - \int_\sigma u_g(\mathbf{y}) \left[\beta G(\mathbf{x}, \mathbf{y})/\alpha + \partial_{n(\mathbf{y})} G(\mathbf{x}, \mathbf{y})\right] d\sigma(\mathbf{y}) =$$
$$= u_f(\mathbf{x}), \qquad \mathbf{x}\in\sigma. \quad (4.31)$$

This is the *boundary integral equation* for $u_g(.)$. Once u_g and the corresponding $\partial_n u$ on the boundary σ are obtained from the equation (4.31), the Green representation (4.30) can be used to calculate the sound field $u(\mathbf{x})$ at any point $\mathbf{x}\in\Omega$ inside the propagation domain Ω. Details about solvability of the BIE (4.31) can be found in (Filippi et al, 1999).

We use boundary equation (4.31) and presentation (4.30) for an applied problem of environmental noise pollution (prediction of sound levels abound a turnpike) in Section 3.

The equations (4.30) and (4.31) describe both the Robin boundary problem ($\alpha=1$, $\beta\neq0$) and the Neumann problem ($\alpha=1$, $\beta=0$). For the Dirichlet problem ($\alpha=0$, $\beta=1$), in a similar manner we obtain the following Green representation and boundary integral equation:

$$u(\mathbf{x}) = u_f(\mathbf{x}) - \int_\sigma \partial_{n(\mathbf{y})} u_g(\mathbf{y}) G(\mathbf{x},\mathbf{y}) d\sigma(\mathbf{y}), \qquad \mathbf{x} \in \Omega, \tag{4.32}$$

$$\int_\sigma \partial_{n(\mathbf{y})} u_g(\mathbf{y}) G(\mathbf{x},\mathbf{y}) d\sigma(\mathbf{y}) = u_f(\mathbf{x}), \qquad \mathbf{x} \in \sigma. \tag{4.33}$$

Historically the first Green representation (4.29) for the acoustic field is the most commonly used. Other possible types of integral representations for the total pressure field of the problem include:

- *Simple layer potential*:

$$u(\mathbf{x}) = u_f(\mathbf{x}) + \int_\sigma \mu(\mathbf{y}) G(\mathbf{x},\mathbf{y}) d\sigma(\mathbf{y}), \quad \mathbf{x} \in \Omega. \tag{4.34}$$

- *Double layer potential*:

$$u(\mathbf{x}) = u_f(\mathbf{x}) - \int_\sigma \mu(\mathbf{y}) \partial_{n(\mathbf{y})} G(\mathbf{x},\mathbf{y}) d\sigma(\mathbf{y}), \quad \mathbf{x} \in \Omega. \tag{4.35}$$

- *Hybrid layer potential*:

$$u(\mathbf{x}) = u_f(\mathbf{x}) - \\ - \int_\sigma \mu(\mathbf{y}) \left[-\partial_{n(\mathbf{y})} G(\mathbf{x},\mathbf{y}) + \gamma G(\mathbf{x},\mathbf{y}) \right] d\sigma(\mathbf{y}), \quad \mathbf{x} \in \Omega. \tag{4.36}$$

Here the unknown layer density μ (**y**) should be chosen to satisfy a boundary condition on the boundary σ. As opposed to the Green representation (4.29), the functions (4.34)-(4.36) are not equal to zero identically outside the propagation domain Ω.

The choice of the most suitable integral representation for a specific applied problem depends on many factors such as geometry, features of the external pressure sources, efficiency of corresponding numeric algorithms, and so on. As shown in (Filippi et al, 1999), the Green presentation (4.29) has some disadvantages when applied to exterior boundary problems. In this case some of the formulas (4.34)-(4.36) may be used.

Boundary problem (4.23)-(4.24) is exterior if the propagation domain Ω is unbounded (extends to the infinity). Then the problem should be supplemented with the Sommerfeld conditions (4.26). The Green representation (4.29) for the exterior boundary problem leads exactly to the same boundary integral equation (4.31) and expression (4.30). It appears that equation (4.31) has a sequence of real eigenvalues (when α/β is real), which means that the solution might be non-unique for some physically possible sources. This leads to instabilities in numerical procedures.

Let us consider the *exterior Dirichlet problem* (4.23)-(4.24) (β=1) and look for a representation of the total pressure field by the *hybrid layer potential* (4.36). Setting the arbitrary parameter $\gamma=i$ in (4.36) produces the following *boundary integral equation* with respect to the unknown function $\mu(\mathbf{x})$ on the boundary σ:

$$-\mu(\mathbf{x})/2+\int_{\sigma}\mu(\mathbf{y})\left[-\partial_{n(y)}G(\mathbf{x},\mathbf{y})+iG(\mathbf{x},\mathbf{y})\right]d\sigma(\mathbf{y})= \\ =u_f(\mathbf{x}),\ \ \mathbf{x}\in\boldsymbol{\sigma}. \tag{4.37}$$

The solution $\mu(\mathbf{x})$ of boundary integral equation (4.37) appears to exist and be unique for any external source functions. Being substituted into representation (4.36), the solution of equation (4.37) determines the pressure field $u(\mathbf{x})$ in the whole propagation domain Ω (it is not zero outside the domain).

However, the hybrid layer representation is not good for all situations because its BIE involves highly singular integrals in the case of the Robin and Neumann boundary conditions.

State-of-the-art application of the BIE method to other three-dimensional PDEs (Lame's equations and Stokes equations) may be found in (Maz'ya, 1989).

1.2 Integral Models Occurring in Physical Problems

Integral models arise in applied scientific or engineering problems when some physical force (property, quantity) is distributed in certain area and its impact on a process under study is not local but accumulated over the area. The accumulation effect can take place in time and space.

The *spatial integral models* describe global situations when certain physical properties or quantities are distributed over a space domain and linked at different points of the domain. Several interesting examples of such systems from transport theory, acoustics, and radiation theory are investigated in the classic book (Morse and Feshbach, 1953). The acoustical example is a membrane fixed in the rigid plate, which is considered below in Section 1.2.1.

The *dynamical integral models* of Section 1.1.1 take into account the *after-effect (persistence, contagion, hereditary effects)* when a continuous sequence of the past states of a dynamical system impacts the future evolution of the system. In a general case, such effects cannot be described by the corresponding differential equations that connect the initial state $x(t_0)$ with its derivatives at the same instant. For example, a simple piecewise kernel $K(t, \tau)=C$ at $t-\tau\leq T$, $K(t, \tau)=0$ at $t-\tau>T$, reduces the linear dynamical VIE (4.6) to a linear *delay differential equation* of the form $x'(t)=F(t)x(t)-F(t-T)x(t-T)$, which is not an ODE.

In mechanics integral dynamical models of elastic persistence in the form (4.4) (where x - deformation, u - strain) were proposed by Boltzman in XIX century. Vito Volterra developed the Boltzman theory and introduced the after-effect concept to other areas, in particular, to population ecology. In general, the after-effect is defined as an arbitrary nonlinear functional of $u(\tau)$, $-\infty < \tau \leq t$. More recently, such integral dynamical models were used in many areas of science and engineering. In Section 1.2.2 we illustrate integral dynamical models with infinite memory used in the stability analysis of nuclear reactors.

1.2.1 Integral Model of Membrane Vibrations.

As pointed out in (Morse and Feshbach, 1953, p.898), "integral equations arise whenever two systems with distributed mass or other relevant parameters are coupled". Let us consider the vibrations of a membrane fixed in the rigid plate. In this case, the coupled systems are the membrane and the air. The vibrations of the membrane raise sound waves that in turn react back upon the membrane, influencing its vibrations. If this effect cannot be neglected, the corresponding equation of the motion will be an integral equation.

Let $Ox_1x_2x_3$ be a system of Cartesian coordinates and the membrane be fixed in the x_1x_2-plane ($x_3=0$). We will denote the displacement of the membrane as $u(x_1, x_2)$ and assume that the motion of the membrane has a simple harmonic nature with the frequency ω, then the membrane velocity in the x_3 direction is $v_n(x_1, x_2) = \partial u/\partial t = -i\omega u(x_1, x_2)$.

Then the velocity potential V in the air medium in the domain $x_1>0$ is described by the following expression

$$V(x_1, x_2, x_3) = \iint_\sigma G(x_1, x_2, x_3, y_1, y_2, 0) v_n(y_1, y_2) dy_1 dy_2, \qquad (4.38)$$

where σ is the membrane area, $G(x_1, x_2, x_3, y_1, y_2, y_3)$ is the *Green's function for the air medium* satisfying the Neumann condition $\partial G/\partial n = 0$ at $x_3=0$. It may be shown that the Green's function has the following form:

$$G(x_1, x_2, x_3, y_1, y_2, y_3) = e^{ikR}/R + e^{ikQ}/Q\,, \qquad (4.39)$$

where

$$R(x_1, x_2, x_3, y_1, y_2, y_3)^2 = (x_1-y_1)^2 + (x_2-y_2)^2 + (x_3-y_3)^2,$$

$$Q(x_1, x_2, x_3, y_1, y_2, y_3)^2 = (x_1+y_1)^2 + (x_2-y_2)^2 + (x_3-y_3)^2.$$

The sound generated in the air domain $x_1>0$ gives rise to an air pressure which now becomes a forcing term in the equation of the membrane motion. The pressure is related to the velocity potential (4.38) by the expression:

$$p = \rho_0 \partial V/\partial t = -i\omega\rho_0 V,$$

where ρ_0 is the density of the air medium in which the sound propagates.

In the considered case of harmonic excitation, the equation of motion of the membrane is the following:

$$\nabla^2 u + k^2 u = -\frac{p}{T}, \qquad k = \omega\sqrt{\frac{\mu}{T}}, \qquad (4.40)$$

where T is the tension, μ is the density (per unit area) of the membrane material. Substituting the expression for the pressure into (4.40), we obtain the following *integral-differential equation* of the sought-for membrane displacement:

$$\nabla^2 u + k^2 u =$$

$$= \frac{\omega^2 \rho_0}{2\pi} \iint_S \frac{e^{ik\sqrt{(x_1-y_1)^2+(x_2-y_2)^2}}}{\sqrt{(x_1-y_1)^2+(x_2-y_2)^2}} u(y_1, y_2) dy_1 dy_2. \tag{4.41}$$

Using the *Green's function for the membrane*, the integral-differential equation (4.41) can be converted to an integral equation.

According to the equation (4.41), the sought-for membrane displacement u depends not only on the value of u at a point on the membrane and neighboring points but also on its values at every point of the membrane in the manner dictated by the integral in (4.40).

This example shows that an integral equation appears when an impulse of one point in a medium may be transmitted to another distant point through a different medium coupled with the first medium. The equation of motion for the first medium contains an additional term arising from the impulse propagation in the second medium. This term will involve all points of the contact surface between the media (the membrane surface in the example above). Another physical example with a similar situation is related to *radiation problems* when the reactive effects of the emitted radiation back on the radiation source cannot be neglected; see (Morse and Feshbach, 1953) for more detail.

1.2.2 Integral Models of Nuclear Reactors Dynamics.

Integral dynamical models are used for simulation of physical processes that take place in nuclear reactors. Here we consider a simple model of such processes adapted from (Podowski, 1986) and (Conduneanu, 1991).

The nuclear reactor power $\rho(t)$ is determined by the flows $\eta_k(t)$ of delayed neurons via the following equations:

$$\dot{\rho}(t) = -\sum_{k=1}^{M} \beta_k \Lambda^{-1}[\rho(t) - \eta_k(t)] - P\Lambda^{-1}[1 + \rho(t)]\nu(t), \tag{4.42}$$

$$\eta_k(t) = \lambda_k \int_{-\infty}^{t} \rho(s)\exp\{-\lambda_k(t-s)\}ds, \qquad k = 1,2,...M, \tag{4.43}$$

where

the constant Λ^{-1} is the mean life time of the instantaneous neurons,

λ_K is the radioactive constant of precursors of delayed neurons in the group of rank k,

β_K is the fraction of delayed terminal neurons in the same group,

M is the number of ranks for delayed neurons,

P is a steady-state value of the reactor power.

The linear scalar functional $\nu(t)$ in (4.42) describes a delayed feedback of circulating reactor fuel. In the simplest case it can be presented as a Voterra operator with infinite memory:

$$\nu(t) = \int_{-\infty}^{t} \gamma(t-s)\rho(s)ds. \tag{4.44}$$

The expressions (4.43) and (4.44) provide clear examples of *integral dynamical model with infinite memory* of the form (4.4). Being substituted into the first equation (4.42), they produce the following nonlinear integral-differential equation with infinite memory

$$\dot{\rho}(t) = -\Lambda^{-1}\left(\sum_{k=1}^{M} \beta_k\right)\rho(t) + \Lambda^{-1}\sum_{k=1}^{M} \beta\lambda_k \int_{-\infty}^{t} \rho(s)\exp\{-\lambda_k(t-s)\}ds - \\ - P\Lambda^{-1}[1 + \rho(t)]\nu(t), \tag{4.45}$$

with respect to the reactor power $\rho(t)$. The only nonlinearity in the equation (4.45) is due to the last term $\rho(t)\,\nu(t)$. Stability of an initial problem for the equation (4.45) was investigated in (Podowski, 1986).

A more flexible model of a controlled nuclear reactor is described in (Conduneanu, 1981) and includes the description of a regulating system of the reactor. In this model, the equation (4.42) for the reactor power $\rho(t)$ remains the same whereas the equations (4.43) and (4.44) are replaced with the following:

$$\dot{\eta}_k(t) = \lambda_k[\rho(t) - \eta_k(t)], \qquad k = 1,2,\ldots M, \tag{4.46}$$

$$v(t) = (c^* x)(t) + \alpha_0(t)\rho(t) + \sum_{j=1}^{\infty} \alpha_j(t)\rho(t - t_j) + \\ + \int_{-\infty}^{t} \gamma(t-s)\rho(s)ds. \tag{4.47}$$

The n–dimensional vector-valued function $x(t)$ in (4.47) represents a state vector of the reactor regulating system, which is described by a traditional integral dynamical model of the automatic control theory:

$$\dot{x}(t) = (Ax)(t) + (b\rho)(t) \tag{4.48}$$

$$(Ax)(t) = A_0 x(t) + \sum_{j=1}^{\infty} A_j x(t - t_j) + \int_{-\infty}^{t} B(t-s)x(s)ds, \tag{4.49}$$

$$(b\xi)(t) = b_0\xi(t) + \sum_{j=1}^{\infty} b_j\xi(t - t_j) + \int_{-\infty}^{t} \beta(t-s)\xi(s)ds. \tag{4.50}$$

The system of integro-differential equations (4.42),(4.46)-(4.50) describes the interaction between the reactor regulating system and the reactor itself. This system is obviously more complex as compared with the model of uncontrolled reactor (4.42)-(4.44). The solutions to the equations (4.42),(4.46)-(4.50) fluctuate around some nominal values representing the steady state of the reactor and its regulating system. It is easy to see that the constant value P of the reactor power $\rho(t)$ corresponds to the constant neuron flows $\eta_K(t)$ and stationary control vector $x(t)$.

Stability of reactor regimes is of obvious practical importance. A stability analysis of the integral-differential equations (4.42), (4.46)-(4.50) is provided in (Conduneanu, 1991). It and includes the stability of the stationary reactor regime and other trajectories close to the stationary one. An interested reader can find more about modelling of nuclear reactors in (Akcasu et al, 1971; Gorjacenko, 1971; Luca, 1979; etc.). Some basic results about the stability of the solutions of nonlinear Volterra integral equations are presented in Chapter 7 in Appendix.

2. MODELLING OF TRAFFIC NOISE PROPAGATION

Applied problems of environmental noise pollution (transportation noise, industrial noise, etc.) are an important part of engineering acoustics (Filippi et al, 1999). The prediction and control of the sound level around industrial noise sources (plant, factory, airport, turnpike, and so on) relies on the analysis of outdoor sound propagation. Such applied problems are quite complicated. They take into account various phenomena such as ground effect, diffraction by obstacles, and propagation in an inhomogeneous medium. A preliminary estimation of the relative importance of each phenomenon is essential.

The required level of applied model accuracy and detalization depends on the essence of an applied problem under study. In particular, the accuracy should not be higher than the accuracy of given experimental data. Often, a precise description of some phenomena is not necessary to obtain a satisfactory answer to our problem.

Here we consider an applied problem of noise propagation due to traffic road (cars, train) over an inhomogeneous surface. The corresponding models should include an investigation of the acoustical characteristics of the ground (the ground effect) and further calculation of the sound field for these characteristics.

Statement of Mathematical Problem. Let us consider a simple model of sound propagation due to traffic road over a plane ground grassy area.

Let $Oxyz$ be a system of Cartesian coordinates (see Figure 4.1). We consider the problem of sound propagation in the half-space $z>0$ from a cylindrical isotropic sound source with the axis parallel to the x-axis. The boundary xy-plane represents the inhomogeneous ground surface that is supposed to be made of a perfectly reflecting straight strip of a constant width (traffic road), which separates two half-planes (two grassy fields) described by the same normal impedance (acoustic resistance).

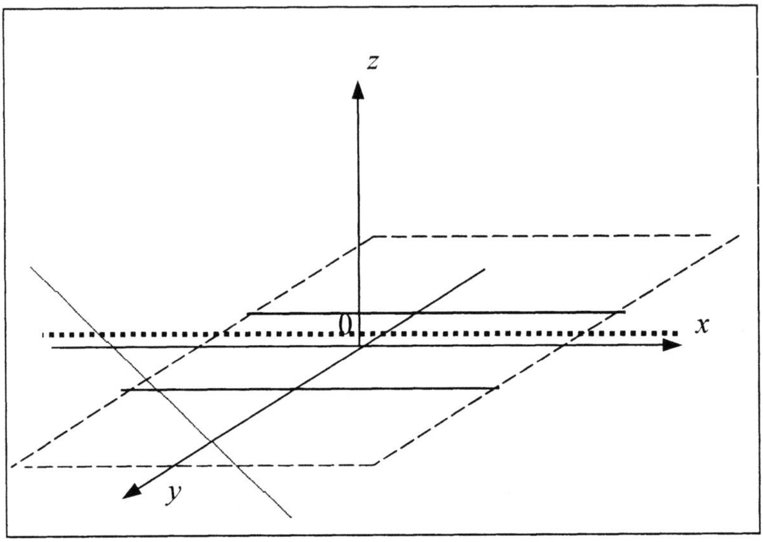

Figure 4.1. The model of sound propagation due to traffic road (the dotted line) over a grassy area.

If we assume that the sound source is infinite in the x direction, then all given and sought-for characteristics of the process of sound propagation do not depend on the coordinate x. Therefore, the described physical system may be modeled by the following two-dimensional boundary problem for the Helmholtz equation:

$$(\Delta + k^2)u(y,z) = \delta(y)\delta(z-z_0), \qquad z>0, \tag{4.51}$$

$$\partial_n u(y,z) = 0, \qquad z=0, \ y\in[a,b], \tag{4.52}$$

$$(\partial_n + ik/\zeta)\, u(y,z) = 0, \qquad z=0, \ y\in(-\infty,a]\cup[b,\infty), \tag{4.53}$$

$$u \sim O(r^{(1-n)/2}), \qquad \partial_n u - iku \sim o(r^{(1-n)/2}), \tag{4.54}$$

where

$[a,b]$ is the road strip,
z_0 is the height of the external sound source (road) above the ground z=0,
∂_n is the normal to the plane z=0,

ζ is the complex impedance of the grassy fields,
k is the wave (frequency) number of the harmonic signal emitted by the sound source.

Since the boundary problem (4.51)-(4.53) is exterior (unbounded), we need to add the Sommerfeld conditions (4.54) for infinitely remote space points.

The boundary surface (xy-plane) is heterogeneous in the problem (4.51)-(4.54). Its internal part (the road strip) is characterized by the Neumann condition and the half-planes are characterized by the Robin condition with the same impedance ζ.

The *impedance* ζ is an important parameter of the ground surface that characterizes the surface ability to adsorb sound pressure. It reflects a so-called *locally reacting model of the surface*, which is approximate but satisfactory for many kinds of surfaces (Filippi et al, 1999) in the sense that it provides an error not greater than measurement errors. The impedance is dependent on the frequency (wave number k) of the emitted harmonic signal and is represented by a complex number.

Method of investigation of the problem (4.51)-(4.54) is based on the BIE method. As opposed to the model examples in Section 1.1.3, we will use the integral representation via the more sophisticated Green's function $G(y,z)$ that satisfies the same impedance condition on the boundary plane z=0:

$$(\Delta + k^2)G(y,z) = \delta(y)\delta(z\text{-}z_0), \qquad z>0, \tag{4.55}$$

$$(\partial_n + ik/\zeta)\, G(y,z) = 0, \qquad z=0, \tag{4.56}$$

$$G \sim O(r^{(1\text{-}n)/2}), \quad \partial_n G - ikG \sim o(r^{(1\text{-}n)/2}). \tag{4.57}$$

The problem (4.55)-(4.57) can be resolved analytically. Different analytical expressions for the Green's function via layer potentials as given in (Filippi et al, 1999), where some practical approximations techniques are also discussed.

Then the Green representation for the solution of problem (4.55)-(4.57) may be written as

$$u(\xi) = G(\xi_0,\xi) + \frac{ik}{\zeta}\int_a^b G(\chi,\xi)u(\chi)d\sigma(\chi), \tag{4.58}$$

where $\xi=(y,z)\in\Omega=\{z>0\}$, $\chi=(y,z)\in\sigma=\{y\in[a,b], z=0\}$, $\xi_0=(0,z_0)$.

The boundary integral equation for $u(.)$ is obtained from the integral representation (4.30) by letting the point $\xi\in\Omega$ tend to a point $\xi_b\in\sigma$ on the boundary $[a,b]$:

$$u(\xi_b) = G(\xi_0,\xi_b) + \frac{ik}{\zeta}\int_a^b G(\chi,\xi_b)u(\chi)d\sigma(\chi). \tag{4.59}$$

It is a one-dimensional linear integral equation with respect to the sought-for function $u(y,0)$, $y\in[a,b]$. Once this equation has been solved, the sound pressure $u(\xi)$ can be calculated anywhere in the half-plane Ω by using the integral representation (4.58).

We shall note that using another Green's function leads to different integral representation and BIE. For example, if instead of $G(y,z)$, we use the Green's function $D(y,z)$ that satisfies the Neumann condition on the boundary plane $z=0$, then the obtained boundary integral equation

$$u(\xi_b) = D(\xi_0,\xi_b) - \frac{ik}{\zeta}\int_{-\infty}^a D(\chi,\xi_b)u(\chi)d\sigma(\chi) - \frac{ik}{\zeta}\int_b^{\infty} D(\chi,\xi_b)u(\chi)d\sigma(\chi), \tag{4.60}$$

includes the integration via an infinite domain. However, the corresponding Green's function is defined by a simpler expression (4.27a).

Obtained results. The major applied results of the model are connected to improved engineering techniques for calculation of an acoustic field generated by traffic roads. Some theoretical results are connected with the estimation of the suitability of the applied model constructed above. Thus, the boundary integral equation (4.58) was investigated numerically using a simple collocation algorithm and the results were compared with the results of an experiment carried out in (Filippi et al, 1999). The results of the numerical simulation fit well the experimental data.

Some generalizations of the model include:

The Case of Additional Obstacles. A traffic road may be surrounded by artificial or natural barriers of various physical character (lines of trees, porous screens, fences, etc.) that protect inhabitants from traffic noise. A natural question arises about the efficiency of such barriers. It requires calculation of the diffraction of acoustic pressure by obstacles of various shape.

The Case of Non-homogeneous Medium. In the above problem, the air was characterized by a constant density and a constant sound speed. In the case of wind or temperature changes, the sound speed becomes a function of space variables. This leads to the Helmholtz equation with variable coefficients.

Another possible cause of heterogeneity is the phenomenon of turbulence. It leads to random differential equations.

The Case of Non-harmonic Sound Sources. In the above model, only harmonic signals were studied. Then it was possible to reduce the wave equation to the Helmgoltz equation. In fact, almost every deterministic acoustic excitation may be considered as a combination of (an integral over) elementary harmonic components. For non-harmonic signals, it is necessary to solve the wave equation of sound propagation directly (see Section 3.2.5 of Chapter 2).

3. MODELLING OF MINE ROPE DYNAMICS

Lifting devices used in mines (mine elevators) belong to complex and expensive engineering systems. Their proper functioning is crucial for safety and mining business efficiency. A possible failure may cause large life and financial losses.

A *mine elevator* consists of a mine rope with a load (cage) on the bottom end of the rope and a powerful electric engine on the surface with a wheel that winds the rope. The device is situated in the shaft of the mine and its size depends on the size of the mine. An average deep mine might have a steel rope of 4 inches in diameter with hundreds yards of length and a mine elevator wheel with a diameter of 4-6 yards.

3.1 Description of Physical Process.

A mathematical description of the mine rope dynamics is the most critical part of applied models of mine-lifting devices. The mine rope is considered as an object of variable length. The dynamics of the objects of variable length requires some generalizations of standard methods of mathematical physics. Such problems do not have eigenvalues and eigenvectors in the common sense because these characteristics become functions of time when the object's length changes. At the first glance, the motion of the objects of variable length represents some kind of non-stationary transition process that does not possess its own modes and movements.

However, numerous observations and experiments show that real-life objects of variable length possess stable forms of movements (modes). For example, one can easily measure several first frequencies (eigenvalues) for longitudinal or transverse elastic vibrations of a mine rope (although these values would be dependent on time) and observe resonance phenomena when these frequencies tend to the frequencies of external forces. The explanation lies in the fact that the speed of the length change is small as compared with the speed of the elastic wave propagation in the object in most engineering systems. For example, the speed of the load lift in mines lies in the interval 10-30 m/sec while the speed of the elastic wave propagation in a steel rope is 4200 m/sec. So, the ratio between these two speeds is less than 0.1 and may be considered as a small parameter (which justifies using various asymptotic methods for solving such problems).

The obvious practical importance of this problem leads to its intensive investigation. Applied engineering techniques for calculating technical parameters of mine elevators (such as the steel rope diameter, the mine wheel diameter, power of the mine elevator engine, and so on) were developed on the basis of static mechanical analysis and approximate dynamic analysis. The dynamics of mine elevators became a "classic example" in the mechanics of the objects of variable length (like simple pendulum vibrations in the classic mechanics of material particle).

The most important dynamical analysis for the mine elevator appears to be the investigation of longitudinal vibrations in the mine rope. Several attempts have been made in solving this problem and the most satisfactory one uses an integral model. It is illustrated below.

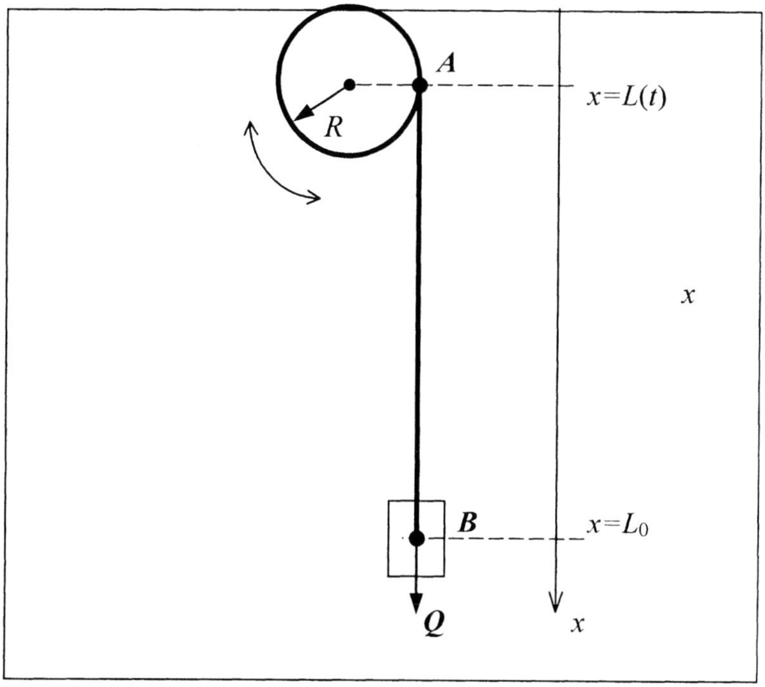

Figure 4.2. The scheme of mine elevator with the load Q.

A natural choice of a coordinate system for this engineering system is the *vertical* axis with the origin in the contact point A of the wheel and the rope (where the rope leaves the wheel, see Figure 4.2). However, the corresponding model of the rope dynamics leads to a boundary problem for partial functional-differential equations. The equations contain the sought-for displacement of the rope together with its value at the boundary point. It caused major difficulty in the boundary problem analysis and this model appeared to be too complicated for practical use.

3.2 Differential Model

The first model of the mine rope dynamics affordable for practice was based on using a *special moving coordinate system* (Goroshko and Savin, 1971). Namely, let us consider the x-coordinate axis directed down from the origin $x=0$ that moves up with the speed $v_C(t)$ (when the load is lifted) or moves down (when the load goes down). Then the contact point A of the wheel and the rope (Figure 4.2) has the variable coordinate $x=l(t)$, where $l(t)$ denotes a variable length of the rope on the wheel rim and may be a given or sought-for function. The bottom end of the rope (the point B) has the constant coordinate $x=L_0$ equal to the total length of the rope. Then the space region of the rope dynamics is $0 \le x \le L_0$.

As usual for small motions in a solid body (Chapter 2), let us choose the displacement (deformation) of the rope $u(x,t)$ as the sought-for variable. The model of the mine rope dynamics in the above coordinate system can be constructed using standard conservation laws (Chapter 2) and is described by the following *one-dimensional wave equation* with respect to the sought-for displacement (deformation) of the rope $u(x,t)$, $0 \le x \le L_0$, $0 \le t$:

$$\frac{q}{g}\frac{\partial^2 u}{\partial t^2} = ES\frac{\partial^2 u}{\partial x^2} + q\left(1 \pm \frac{v_C}{g}\right), \tag{4.61}$$

where

$q=\rho S$ is the weight of a unit length of the rope,
ρ is the mass density,
S is the cross-section area of the rope,
g is the acceleration of gravity,
E is the module of elasticity.

The sign $\pm$ in the right-hand part of equation (4.61) depends on the direction of external force (whether the cargo goes up or down).

The boundary condition at the bottom end $x=L_0$ has the following common form for a free end with a joint mass (compare with (2.29c)):

$$Q/g\partial^2 u(L_0,t)/\partial t^2 + E\partial u(L_0,t)/\partial x = 0 \tag{4.62}$$

where Q is the weight of the lifting load (cargo). If the load weight Q is negligible as compared with the rope weight qL_0, then we can use the free end boundary condition:

$$\partial u(L_0,t)/\partial x = 0.$$

The boundary condition at the up end $x=l(t)$ appears to have the following non-integrable form:

$$u[l(t),t)] = \int_0^t \left(\partial u(l(\xi),\xi)/\partial x\right)\dot{l}(\xi)d\xi. \tag{4.63}$$

and contains the unknown function $u\ (l(t),\ t)$ in the integrand expression.

So, in the new moving coordinate x, the complexity of the problem is shifted from the motion equation (which becomes a common wave equation) and the bottom end boundary condition (which is now fixed at $x=L_0$) to the up end boundary condition at $x=l(t)$.

The law $l=l(t)$ of the load lifting (or falling) is determined by the dynamics of the mine engine and may be a given or sought-for function (depending on some assumptions about the engine and rope parameters). Namely, in a general case the variable length of the rope $l(t)$ is determined as

$$l(t) = l(0) + \int_0^t \left(1 + \partial u(l(\xi),\xi)/\partial x\right)^{-1} v_c(\xi)d\xi. \tag{4.64}$$

where $v_c(t)$ is the speed of the load lift (and of the coordinate origin $x=0$).

In many specific cases of small displacements the dependence on the $\partial u(L_0,t)/\partial x$ may be omitted and the variable length of the rope $l(t)$ is approximately equal to

$$l(t) = l(0) + \int_0^t v_c(t)dt. \tag{4.65}$$

The boundary problem (4.61)-(4.64) is considered under the following common initial conditions :

$$u(x,0) = u_1(x), \qquad \partial u(x,0)/\partial t = u_2(x). \tag{4.66}$$

Let us denote the rotation (angle displacement) of the mine engine as $\varphi=\varphi(t)$. Then the linear velocity of the point A on Figure 4.2 is $v_C(t)=Rd\varphi(t)/dt$.

There are two main cases in the mine engine dynamics:

Case A (case of the given variable length). The mine engine is powerful enough and its dynamics does not depend on the dynamics of the mine rope. Then the rotation $\varphi=\varphi(t)$ of the mine engine is assumed to be a given function in the problem (4.61)-(4.66).

Case B (case of given external force). The mine engine dynamics depends on the dynamics of the mine rope. Then the rotation $\varphi=\varphi(t)$ of the mine engine is determined by the following ordinary differential equation:

$$\frac{d}{dt}\left(J\frac{d\varphi}{dt}\right) = M(t) - RES\frac{\partial u(l,t)}{\partial x}, \tag{4.67}$$

where

J is the mass moment of inertia of the mine wheel,
R is the radius of the wheel module of elasticity,
$M(t)$ is the (given) external torque developed by the engine (Vierck, 1969).

The problem (4.61)-(4.66) in case A may be considered as a linear problem under the given speed $v_C(t)$ of load lift and the simplified assumption (4.65). The boundary problem (4.61)-(4.67) in the case B under the given torque $M(t)$ is always nonlinear in virtue of the functional relationship of the sought-for functions $u(l(t), t)$ and $l(t)$ in the boundary point $x=l(t)$.

The main difficulty in analysis of the nonlinear boundary problem (4.61)-(4.67) is connected with handling the non-classic, non-integrable boundary condition at $x=l(t)$. As it was discussed in the previous section, a common technique for such complicated case is transferring the problem to integral equations. The corresponding integral methods were developed in (Goroshko and Savin, 1971) and sequential works.

3.3 Integral Model

Using the direct integration by the spatial coordinate x and excluding the boundary conditions, the boundary problem (4.61)-(4.67) for the wave equation (4.61) with the initial functions $u_1=u_2\equiv 0$ is reduced to the following *integral equation of the second kind* with respect to the sought-for displacement (deformation) of the rope $u(x,t)$, $0\leq x\leq L_0$, $t\geq 0$:

$$u(x,t) = -\int_{l(t)}^{L_0} K(x,s,l(t))\frac{\rho(s)}{g}\left(\frac{\partial^2 u}{\partial t^2} - g \mp \frac{dv_C}{dt}\right)ds - \frac{Q}{g}\frac{L_0 - l(t)}{ES}\left(\frac{\partial^2 u(l(t),t)}{\partial t^2} - g \mp \frac{dv_C}{dt}\right) + \int_0^t \frac{\partial u(l(\xi),\xi)}{\partial x}\frac{dl(\xi)}{d\xi}d\xi. \tag{4.68}$$

The kernel of the integral-differential equation (4.68) is symmetric with respect to the coordinates x, s:

$$K(x,s,l) = \frac{\min(s,x)-l}{ES} = \begin{cases} (s\text{-}l)/ES, & s \le x, \\ (x\text{-}l)/ES, & s \le x. \end{cases} \tag{4.69}$$

After some transformations, the integral-differential equation (4.68) takes the form

$$U(x,t) = -\int_{l(t)}^{L_0} K(x,s,l(t))\frac{\tilde{\rho}(s)}{g}\times \\ \times\left[\frac{\partial^2 U}{\partial t^2} + \frac{d}{dt}\left(\frac{\partial U(l(t),t)}{\partial x}\frac{dl(t)}{dt}\right) - g \mp \frac{dv_C}{dt}\right]ds. \tag{4.70}$$

with respect to the function $U(x,t) = u(x,t) - \int_0^t \frac{\partial u(l(\xi),\xi)}{\partial x}\frac{dl(\xi)}{d\xi}d\xi.$

The limits of integration and the kernel in the non-linear integral equations (4.68) and (4.70) depend on the given (or sought-for) function $l(t)$. This function is determined with the formula (4.64) by the given load lifting speed $v_C(t)$ in case A or by the engine dynamics equation (4.67) in case B.

If we assume that the speed $v_C(t)$ of load lift is given (case A) and the simplified assumption (4.65) is valid, then

$$dl(t)/dt = v_C(t) = Rd\varphi(t)/dt.$$

is given and we can rewrite the equation (4.69) as

$$U(x,t) = -\int_{R\varphi(t)}^{L_0} K(x,s,R\varphi(t))\frac{\tilde{\rho}(s)}{g}\times \\ \times\left[\frac{\partial^2 U}{\partial t^2} + R\frac{d}{dt}\left(\frac{\partial U(R\varphi(t),t)}{\partial x}\frac{dR\varphi(t)}{dt}\right) - g \mp R\frac{d^2\varphi}{dt^2}\right]ds. \tag{4.71}$$

where the function $\varphi(t)$ is given (in case A) or defined by the following equation (in case B):

$$\frac{d}{dt}\left(J\frac{d\varphi}{dt}\right) = M(t) - RES\frac{\partial U(R\varphi(t),t)}{\partial x}. \tag{4.72}$$

In the special case when the function $l(t)$ in the lower limit of integration is constant (a motionless load), the integral-differential equation (4.71) is a linear Fredholm-type integral equation of the second kind with respect to the sought-for function $U(x,t)$, $0 \leq x \leq L_0$, $t \geq 0$.

In case B, the system of equations (4.71)-(4.72) in two sought-for functions $U(x,t)$, $\varphi(t)$, $0 \leq x \leq L_0$, $t \geq 0$, under the given torque $M(t)$ is *always nonlinear* because of the functional relationship $u(R\varphi(t),t)$. In this case, we obtain a nonlinear system of integral-differential equations (4.71)-(4.72) *with a sought-for function in the limit of integration.* Similar problems arise also in other applications; some of them are considered in the Chapters 5 and 6.

Methods of investigation. Various approximate methods (analytic and numeric) were applied to the integral-differential equation (4.71) in both cases A and B. They include the method of momentum, Galerkin method, an iteration method, linearization by an averaging algorithm, various asymptotic methods of nonlinear mechanics, and so on.

As mentioned in Section 2.1, the ratio of the cargo lift speed to the elastic wave speed is less than 0.1 in a steel rope. So, this ratio may be considered as a small parameter (which justifies the use of asymptotic methods for solving such problems).

Obtained results. The major applied results of the model are related to improving engineering techniques for calculation of technical parameters of mine elevators. Some theoretical results include the following:

- The vibration energy and stress in a perfect elastic string increase when the length of the string decreases.
- The amplitude and frequency of the first (fundamental) vibration mode for a perfect elastic string increase when the length of the string decreases. It leads to resonance behavior when the length of the mine rope tends to zero (in the up point of the lift).
- The motion of the objects of variable length is characterized by *two groups of standing waves.* These groups have the same set of wave frequencies but distinct waveforms and wave phases. They represent two fundamental systems of eigenvectors corresponding to the same real eigenvalues.
- A similarity is established between the dynamic problems for the objects of variable length and the dynamic problems for elastic objects carrying a moving inertial load (like an elastic hose with a fluid flow inside). The

common characteristic feature of these two applications is the occurrence of two groups of standing waves (see the above).

3.4 Some Generalizations.

The exposed results have been generalized for more realistic real-life cases in different directions.

1. Energy dissipation. If we consider a viscoelastic mine rope, then the stress $T(x,t)$ is connected with the rope displacement $u(x,t)$ as

$$T(x,t) = EF\partial u/\partial x + \alpha\partial u^2/\partial x\partial t,$$

where α is a coefficient of the energy dissipation because of the elastic imperfectness of the rope. Then the differential equation (4.61) and integral equation (4.68) of the rope dynamics include an additional term $\alpha\partial u^3/\partial x^2\partial t$,

Including the energy dissipation into the model made it possible to determine a critical speed that appears to depend on the parameter α of the energy dissipation. Similarly, the boundary condition (4.63) for the load dynamics can include a possible resistance force $F_R(t)$ of the air medium.

2. Endogenous mass moment of inertia of the engine wheel. If the total mass qL_0 of the mine rope is comparable to the mass of the mine engine wheel, then the wheel mass moment of inertia J in the equations (4.67) and (4.72) of engine dynamics may depend on the mass $ql(t)$ of the rope winded on the wheel (this value belongs to sought-for functions of the model) as

$$J = J(l) = Rql(t)/g = R^2 q\varphi(t)/g,$$

This expression delivers an additional nonlinearity to the nonlinear system of equations (4.71)-(4.72).

3. Sliding of the rope at the wheel contact point. Another important phenomenon is the so-called *sliding of the mine rope* at the contact point of the mine wheel and rope (point A on Figure 4.2). The physical cause for the sliding is that the friction forces on the wheel rim are not enough to keep the elongated rope. This phenomenon has been discovered experimentally and appears to be of practical importance. Then the boundary condition (4.63) at the up end $x=l(t)$ has more complicated form:

$$u[l(t),t)] = \int_0^t \frac{\partial u(l(\xi),\xi)}{\partial x}\left(\dot{l}(\xi) - \Delta\dot{l}(\xi)\right)d\xi + \Delta\dot{l}(t)\frac{\partial u(l(t),t)}{\partial x}, \qquad (4.73)$$

where "the sliding path" $\Delta l(t)$, $0 \le \Delta l(t) < l(t)$, is a function of time determined via a specific empirical formula. Then, one can find that after transformations a similar term appears in the integral-differential equation equations (4.68) and (4.71). An analysis of the sliding impact on the rope dynamics was provided in (Goroshko and Savin, 1971).

4. Transverse and torsional vibrations. The most essential part in the dynamical analysis of mine-lifting devices appears to be the investigation of longitudinal vibrations in the mine rope. The joint influence of *transverse vibrations* and *torsional vibrations* of the rope represents another interesting applied problem. The corresponding models have been constructed and their analysis has shown that these vibrations add small corrections to the total maximum dynamic values of stress in the rope based on longitudinal vibrations analysis.

Chapter 5

Modelling in Bioengineering

Bioengineering is a rapidly growing field that has been flourishing during the last decades. It involves a wide diversity of problems ranging from physiological processes, including blood circulation in the body, gas exchange in the lungs, control of cell volume, and metabolic and protein design, to robotics and transport phenomena in biological systems, all of which require bioengineering methods. Bioengineering represents the full collaboration and integration of approaches of Engineering, Biological Sciences, and Mathematical Modelling.

From the molecular and cellular levels up to the level of the organism biology is becoming more accessible to mathematical approaches. The accelerating pace of discoveries in the Biological Sciences brings more questions for research. On the other hand, it requires engineering design and mathematical modelling to explain the phenomena and solve the resulting problems.

Engineering develops better analysis and description of complex biological systems and it creates new biological structures and instrumentation for research and clinical applications as well as new types of man-made systems. Medicines created with the help of bioengineering become smarter, and targeting disease more effectively than traditional medicines. They use more rational and reasonable recourses to develop new drugs or vaccines and improve existing ones. The authors will provide just a few examples. Insulin that is so important for diabetics can come from the pancreas of a pig or cow. Because pig insulin is nearly identical to human insulin, pigs are the most common source, but some diabetics react inadequately to pig insulin. In addition, it is difficult to obtain the required quantity of insulin when pigs are the main source. Genetic engineering has found another way to produce large amounts of insulin. Using recombinant DNA technology, the gene for producing insulin is inserted into the Escherichia coli (E. coli) bacteria. These genetically engineered bacteria are turned into tiny insulin-producing factories. The resulting insulin is so close to human insulin that it is virtually impossible to distinguish one from the other. In another example, sheep have been genetically modified to express the human blood clotting IX factor in their milk. This protein is important to hemophiliacs, whose bodies do not produce the protein because of a genetic mutation.

Engineering in Biological Sciences applications uses principles of *synthesis* of new biomaterials and biosystems together with *analysis* of the physical and engineering principles used by nature in biological systems.

Mathematical models and methods are becoming increasingly significant and give an in-depth account of the practical use of mathematics in many important and diverse areas of bioengineering and biosciences (Murray, 1989; Hoppensteadt and Peskin, 1992; and others). Some mathematical models have been developed pretty well, and they offer a good description of a biosystem, helping to understand it and produce man-made analogues. Other biosystems, however, have seen just the first steps toward the development of mathematical models.

The mathematical methods for biological systems and processes are based on a wide range of mathematical tools such as ordinary and partial differential equations, difference and integral equations, and stochastic and deterministic models. They can involve elements of stochastic and randomness. The same models may be applied across a range of bioengineering problems. To create an accurate model, it is important to know the advantages and disadvantages of mathematical methods. The choice of model should be also conditioned by the modelling goals and the problem's complexity.

The purpose of this Chapter is to show the variety of basic mathematical tools that have wide use in bioengineering applications. Such models, especially their modifications and combinations, provide a good description to biological system or process, helping to create valid bioengineering models and devices. The first theoretical models are relatively simple and represent just the most important mechanisms or features of the situation under study. Once the simplest case is understood, the model's shortcomings usually become clear, and more realistic assumptions can be added.

Section 1 is devoted to a description of several basic models that use various mathematical tools. All model structures are very clear, and all effects are easy to explain. Most of these models have been elaborated and improved, generalized and combined to create new models that better describe this or that specific process. For example, combining the logistic model of Section 1.1.2 with the diffusion model of Section 1.6.2 leads to the well-known Fisher population model, confirmed in many applications. This section shows the connections between different models illustrating model changes after adding some new characteristics and constrains.

Section 1 discusses the most important characteristics of models (stationary points, stability, and bifurcation points); and, in addition, it introduces a new nonlinear integral model (that will be investigated in Chapter 2).

Section 2 contains some interesting questions related to the qualitative and bifurcation analysis of models. Such analysis lies in finding the *oscillatory and stationary solutions* for the models and the *bifurcational values* of parameters at which a new stationary solution appears or another significant change in behavior takes place.

Bifurcations in the difference and differential models have been thoroughly studied in many publications during the last two decades. Although the dynamics of such models is fairly complicated, some specific patterns have been discovered for particular types of the models. The integral equations models are more general than the difference and differential equations models and are able to describe various patterns of population behavior. They have many unsolved questions and open issues. In light of the above, the authors have chosen a nonlinear integral model and provided its detailed bifurcation and stability analysis. An initial analysis shows the presence of new bifurcations and oscillatory regimes in the nonlinear integral model (for larger values of the bifurcation parameter), which are analogous to phenomena known for the difference models of population dynamics (May, 1976). The connection between the integral and the difference models as well as other possible bifurcations in the model under study are discussed in Section 2.3. Section 2.4 treats some residual problems.

1. MODELS OF POPULATION DYNAMICS AND CONTROL

For many reasons it is important to study how populations of living creatures change and their possible impact on human lives. Will a population grow to a point that is not sustainable? Likewise, variations in the size of non-human populations have relevance to humans. Once-abundant fish species no longer exist in sufficient numbers to be harvested for food. Populations of endangered animals continue to be affected by human interventions. Controlled growth of bacterial populations is essential in many bioengineering manufacturing processes.

The realistic study of populations involves various specific aspects including data on birth and death rates, food supplies, migration, social behaviors, genetics, the interaction of species with their environments and among themselves, etc. An effective mathematical description contributes significantly to quantitative understanding of how populations change in time.

This chapter introduces some mathematical and computational tools needed to describe quantitative change of populations in time. The first considered models involve the exponential and logistic equations. Then we move from age-dependent models and models with intra-species competition to delay and difference equation models and, finally, to the models of diffusion processes. The goal is not to discuss all existing models, but to show their basic characteristics using examples of fundamental mathematical equations applied to bioengineering.

1.1 Classic Models for One Species Population

In this section, the authors discuss the exponential and logistic models of a single population that describe a few of the most important population characteristics. The major advantage of these models is their simplicity. Their extensive study has resulted in improvement and generalization by the addition of different features and more realistic assumptions.

1.1.1 Malthus Model

The dynamics of a one-species population under ideal conditions is described by the following *conservation equation*:

$$dN/dt = \text{births} - \text{deaths} + \text{migration} \tag{5.1}$$

where $N(t)$ is the number of individuals or the *population size* at time t.

The simplest form of (5.1)

$$dN/dt = \mu N \tag{5.2}$$

with no migration and the birth and death rates proportional to $N(t)$ is the first well-known mathematical description of a one-species population called the *Malthus model* (T.Malthus, 1802) although it was first suggested by Euler in 1760. In this model $\mu=b-d$ is the *coefficient of population intrinsic growth* or the *net birth rate*; b is the *birth-rate coefficient*, d is the *death-rate coefficient*.

The equation (5.2) has the exponential solution

$$N(t) = N_0 e^{\mu t} \tag{5.3}$$

where $N_0 = N(0)$ is the initial size of population. Analyzing the solution (5.3), we conclude that the population grows exponentially if the birth rate exceeds the death rate, $\mu > 0$ (this case is represented with a gray curve in Figure 5.1), and dies out (faces extinction) if $\mu < 0$.

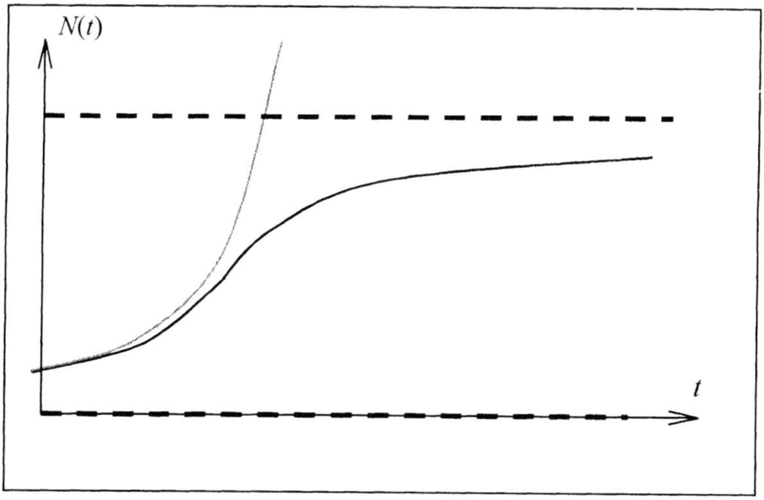

Figure 5.1. Exponential (in gray; $\mu > 0$) and logistic (in black) solutions. The dashed lines represent two stationary trajectories of the logistic curve.

In reality, this model is acceptable only for an unlimited food resource, no competition, and a large population size $N(t)$. Only a few species meet these requirements for a sufficiently large time interval. Let us take the Asian long-horned beetle. It does not have natural predators, and has unlimited food resources in the US – hardwoods such as maple, poplar, birch, elm, ash, horse chestnut, and willow. These beetles spend most of its life deep inside a tree. Unlike most other beetles, which eat dead or dying trees, this insect prefers healthy wood and eats its way through the wood, disrupting the nutrient system of the tree and leaving it vulnerable to rot and pests. Brought to the U.S. in wooden packing crates from China and Hong Kong, Asian long-horned beetles turned up in Brooklyn, N.Y., in the mid 1990s, and in Chicago two years later. They have already destroyed thousands of trees. If they get loose in the country, damage could exceed $650 billions. It's the greatest thread to US forestry since the gypsy moth.

Another example is related to the population of fire ants accidentally introduced to the southern United States from Brazil in the 1940s. With absence of competition (their native predator, anteater, cannot be adapted to US) and plenty of food and space the fire ants spread really fast across the southeastern United States damaging natural ecosystems, destroying crops, threatening brown pelicans, reptiles and amphibians, reducing the number of native ant and other species.

1.1.2 Verhulst-Pearl Model

A population can grow exponentially in accordance to the Malthus model (5.2) during a short period of time, when its size is far from its limits of growth. However, when the population reaches its limit, its size can fluctuate, even chaotically. It means that there must be some additional factors that affect the population growth in the long run. Let us discuss some improvement of model (5.2) that leads to another very famous and more realistic model of a single-species population.

The birth and death rates cannot be constant for a very long time. In a real environment, the increase of a population initiates the shortage of required resources; competition in the population and between different populations, etc. Higher population densities make it easier for a disease to spread out. As a result, the birth rate decreases and the death rate increases. It is reasonable to assume that as a population density increases and approaches the carrying capacity of its environment, the birth rate will decline and the death rate will increase. Then the net rate $\mu = b_0 - d_0$ at some initial moment t_0 with initial growth and death rates of b_0, d_0 . In assumption that the birth and death rates are linearly related to the size of the population N, the birth and death rates can be redefined as

$$b = b_0 - k_b N, \qquad d = d_0 + k_d N, \tag{5.4}$$

where k_b is the rate at which the birth rate deccreases and k_d is the rate at which the death rate increases as N grows. The population size N will stabilize, if $b = d$, or, taking into account (5.4), the population size reaches $\mu /(k_b + k_d)$, usually denoted by K and called a *carrying capacity* or *environment capacity* of the population. This constant represents the medium resistance for population growth. Now, substituting (5.4) into the original conservation equation (5.1) and providing simple mathematics, we get:

$$dN/dt=\mu N(1-N/K). \qquad (5.5)$$

The factor $(1-N/K)$ is negligible when the size of a population is small compared to K and essential with N increases.

The model of a single-species population (5.5) was first proposed in 1838 by the Belgian mathematician Pierre-François Verhulst, and rediscovered in 1920 by Raymond Pearl and Lowell Reed, who applied it to various biological problems ranging from modelling of populations of fruit flies and yeast cells grown under controlled laboratory conditions to modelling of the population of the United States.

The analytical solution of the Verhulst model (5.5)

$$N(t) = K/(1+Ce^{-\mu t}),$$

is known as the *logistic curve of growth*, and depictured in Figure 5.1. In the solution, $C = K/ N_0 -1$ =const.

Stationary trajectories (*equilibrium states* or *steady states*) of the model (5.5) as well of any differential equation

$$dN/dt = f(N),$$

are solutions of $f(N) = 0$. Here $f(N)$ is a nonlinear function of N . To check the stability of a stationary trajectory N^*, we have to take a look at linearization of the original differential equation about $(N-N^*)$, e.g. consider small perturbations $N(t)-N^*$ such that $(N-N^*)^2$ neglected compared with $|N-N^*|$. In general, stationary trajectories N^* of $dN/dt = f(N)$, are linearly stable with respect to the small perturbations if $f'(N^*) < 0$, and unstable if $f'(N^*) > 0$.

The Verhulst-Pearl model (5.5) has two *stationary trajectories*: $N_0^*=0$ and $N_1^*=K$. Moreover, the trivial solution $N_0^*=0$ is unstable and a trivial solution $N_1^*=K$ is stable i.e. $N(t)\to K$ at $t\to\infty$. In other words, the population grows exponentially from any small increment of the initial state $N(0)=0$, and tends toward the size K in the long run. Theoretically, when N reaches the environment capacity K, equilibrium will be reached and the population will be stable. If N_0 is less than K, then $C>0$ and $N(t) = K/(1+Ce^{-\mu t}) \to K^-$, as $t\to\infty$, therefore the rate of growth is positive and the size of population will increase toward the carrying capacity. If N_0 exceeds K, then $N(t) = K/(1+Ce^{-\mu t}) \to K^+$, as $t\to\infty$, the rate becomes negative and the population will decrease toward K. It reflects the fact that when the population grows essentially resources are reduced, the population size is bounded and tends to some value K that determines the size of the stable steady state population in a given habitat at the absence of other populations; and μ measures the rate at which it reaches.

This simple non-linear Verhulst-Pearl model (5.5) has appeared to be good enough for many populations of microorganisms, plants and animals. The parameter K has a clear biological meaning for populations with a strong interaction among individuals that control their reproduction. For example, rodents have social structure that controls reproduction, birds have territoriality, and plants compete for space and light. However, the parameter K has no clear meaning for organisms whose population dynamics is determined by the balance of reproduction and mortality processes (e.g., most insect populations). In this case, the equilibrium population density does not necessarily correspond to the amount of resources. Thus, the term "carrying capacity" becomes confusing.

1.1.3 Population Control and Harvesting

What is the largest sustained yield that a certain population will produce if other species or the same species of different ages are added to the given area? How much fish can be harvested without harming the fish population? What is the environment capacity of the population under study? What is the maximum yield with the minimum effort? Which trees should be selected for harvesting? When should crops be gathered to keep their flavor, taste and nutrition? How much time does it take for a pest (i.e., European cornborer) to adjust to a new type of pesticides? These and similar questions should be investigated before providing harvesting. We have to be able to make predictions and forecast for many real-life situations involving biological self-reproduction in human-controlled environment.

Bioengineering is constantly working on inventing new products. Genetic engineering allows us to take genes from just about anything, and splice them into a genetic makes up, to generate new features of products, like disease-resistant, pest-resistant, or more nutritious crops, or tomatoes grown in cold climates because it contains a gene from a cold-resistant fish.

Harvesting as a special type of a population control is one of the most important processes in the real world. We harvest various crops, animals, trees, develop different bioproducts, mine fossil fuels etc to survive. Unreasonable and irrational harvesting may lead to exhaustion and even disappearance of natural resources. The passenger pigeon, the most abundant bird in North America in 1800s, became extinct by 1914 due to overhunting. Elephants and rhinoceroses are threatened in Africa because of their trunks and horns. Scientific and controlled harvesting minimizes deleterious effects for a population or does not harm it at all. A few species of 110 Hawaiian lobelias survive only because scientists artificially pollinate them.

To illustrate the population control, let us modify logistic model (5.5) by adding harvesting to it. We will assume that the harvesting intensity (number of lines or nets in the water for fishing, number of traps in a given area for trapping, etc.) is proportional to the population size $N(t)$ with the harvesting coefficient η that measures the intensity/effectiveness of the harvesting method applied. To produce a more reasonable model, the parameter η should also reflect the total cost of harvesting resources (e.g., labor and capital employed in harvesting including the actual costs of fuel, gear, and supplies, as well wages, interest and depreciation on capital).

We assume that each member of the population has an equal nonzero probability of being harvested at any moment. Then, the model (5.5) with harvesting turns to be:

$$dN/dt=\mu N(1-N/K) - \eta N, \tag{5.6}$$

Introducing $\mu' = \mu - \eta$ and $K' = K(\mu - \eta)/\mu$ to the last model we note that (5.6) is rewritten as

$$dN/dt=\mu' N(1-N/K').$$

Thus, if a population follows the logistic model (5.5) in the absence of harvesting, then it continues to follow a logistic model being subjected to harvesting, but with a lower net growth rate and lower carrying capacity. It means that the behavior of model (5.6), its basic results and characteristics such as stationary solutions and stable solutions, can be investigated as for model (5.5). For example, if $\mu'>0$, model (5.6) has two stationary solutions: unstable $N_0^*=0$ and stable $N_1^*=K'$, a yield of harvesting can be found from

$$Y(\eta)=\eta N_1^* = \eta K(\mu - \eta)/\mu.$$

We would like to note, that for sufficiently intense harvesting we might get a negative effective net growth rate μ' even for a positive net growth rate μ. It will cause the extinction of the population because of the "*overharvesting*".

1.2 Age-Dependent Models for One Species Population

The models of population dynamics based on ordinary differential equations (for example, models (5.2),(5.5)) are valid only for long time intervals as compared with the longevity of individuals. Such models do not provide an explanation for many experimentally observed phenomena whose duration is of the same order of magnitude as the longevity of an individual. The fertility and mortality of individuals depend on their age; individuals of age t can produce descendants only when t is from a certain interval specific for each ecological system. To describe such effects, it is necessary to take the population age into account thought it is more difficult to deal with age structure than with the total size of a population.

Population age-distribution is one of the key endogenous intrinsic factors influencing the population dynamics and one of the most crucial biological components. Many modifications of basic age-dependent models of ecological systems have been developed during the last three decades. Among various mathematical tools used for description of age structure of biological populations are integral equations or partial differential equations. One of the key advantages of integral models over differential ones is the *reduction of the dimension of the sought-for function* that simplifies analytical and numerical analysis of the models.

In this section, we study integral and differential models of a population description and an interrelation between them.

1.2.1 Linear Integral Model (Lotka Model)

One of the first models that describe the dynamics of age-distributed single-species population in stationary environment under unlimited food resources was suggested in (Sharpe & Lotka, 1911). This model is presented by the linear *integral renewal equation* (see also (Webb, 1985; Poluektov et al, 1980) :

$$X(t) = X^{-}(t) + \int_0^t m(\tau)l(\tau)X(t-\tau)d\tau\,, \quad t \geq 0, \tag{5.7}$$

where

$$X^{-}(t) = \int_t^T m(\tau)l(\tau)\phi(\tau - t)d\tau\,, \quad 0 \leq t \leq T.$$

Here

- $X(t)$ is the *birth-rate intensity* (the number of individuals born per time unit at time t),
- $m(\tau)$ is the *age-specific fertility rate* (the average number of individuals produced per time unit by one individual of age τ),
- $l(\tau)$ is the *age-specific fatigue function* (the specific part of the individuals surviving to age τ),
- $\phi(\tau)$ is the initial distribution (at t=0) of the individuals with respect to their age τ,
- T is the maximum age.

Providing the substitution $t-\tau\rightarrow\xi$, the *integral model* of population dynamics (5.7) may be written down in the form of

$$X(t) = \int_{-T}^{t} m(t-\xi)l(t-\xi)X(\xi)d\xi \tag{5.8}$$

with the initial condition of

$$X(\xi)=\phi(-\xi),\ \xi \in [-T,0].$$

1.2.2 Linear Differential Model (Lotka–Von Foerster Model)

The age structure of biological populations can also be defined using partial differential equations. Assuming a stationary environment, the differential age-dependent model of population dynamics analogous to the linear integral model (5.7) is described by the following balance relation (Gurtin & Maccamy, 1974, Streifer, 1974, Swick, 1977, Webb, 1985):

$$dx(\tau,t) = \partial x/\partial \tau \, d\tau + \partial x/\partial t \, dt = -d(\tau)\, x(\tau,t)\, dt$$

where

$x(\tau,t)$ is the *population density age distribution* at time t of the age range τ to $\tau + d\tau$ (the number of individuals of age τ at time t);

$d(\tau)$ is the *age-specific death rate* (a specific part of the individuals of age τ died per time unit);

$d\tau/dt = 1$ since τ is chronological age;

$d(\tau)x(\tau,t)dt$ is the number of the population of age τ that dies in a small interval of time dt;

$\partial x/\partial \tau \, d\tau$ is the contribution to the change in $x(\tau,t)$ with age increase.

Dividing the last equation by dt we get

$$\partial x/\partial \tau + \partial x/\partial t = -d(\tau)\, x(\tau,t), \qquad (5.9)$$

The last equation is called *the evolutionary equation.* The boundary condition

$$x(0,t) = \int_0^T m(\tau)x(\tau,t)d\tau \qquad (5.10)$$

is called *the fertility equation* for obvious reason. The initial condition

$$x(\tau,0)=\phi(\tau), \ \tau\in[0,T],$$

states that the population has a known age distribution $\phi(\tau)$.

The content of the functions $m(\tau)$ and $\phi(\tau)$ and the parameter T is the same as in the integral model (5.7). Model (5.9)-(5.10) is known as the *Lotka-Von Foerster model.* The *population size* (the total number of individuals) at time t is obtained in model (5.9)-(5.10) by adding up all individuals between the ages 0 and T:

$$N(t) = \int_0^T x(\tau, t) d\tau$$

It was shown in (Poluektov et al, 1980) that the solution x of model (5.9)-(5.10) contains oscillatory components whose maximum period does not exceed twice the duration of reproduction age. In the case of single breeding, the maximum period is equal to the reproduction age, and the oscillations do not decrease.

1.2.3 Equivalence of Integral and Differential Models

Let us illustrate the interrelation between integral population models and their differential analogs.

In the PDE model (5.9)-(5.10), the sought-for function $x(\tau, t)$ depends on the individual age τ and current time t. After replacing the age τ with the instant $\xi = t - \tau$ of the individual birth and introducting the function $\bar{x}(\xi, t) = x(t - \xi, t)$, equation (5.9) becomes:

$$\partial \bar{x} / \partial t = -d(t - \xi) \, \bar{x}(\xi, t)$$

and its analytical solution is:

$$\bar{x}(\xi, t) = \bar{x}(\xi, \xi) \exp(-\int_\xi^t d(\tau - \xi) d\tau). \tag{5.11}$$

Substituting (5.11) into (5.10) and introducing a new one-dimensional sought-for function

$$X(t) = \bar{x}(t, t) = x(0, t),$$

we obtain the following integral model:

$$X(t) = \int_{t-T}^{t} m(t - \xi) e^{-\int_0^{t-\xi} d(v) dv} X(\xi) d\xi \tag{5.12}$$

which is equivalent to the partial differential model (5.9)-(5.10) and has the same form as the integral model (5.8). Hence, the fatigue function $l(\tau)$ in Lotka integral model (5.7) is related to the death rate $d(\tau)$ by the formula:

$$l(\tau) = \exp(-\int_0^\tau d(\xi)d\xi). \tag{5.13}$$

Similar transformation from a differential model to an integral model is also possible in more general cases when the following population factors are taken into account: a non-stationary environment (the functions $m(\tau,t)$, $d(\tau,t)$ and $l(\tau,t)$ depend explicitly on the time t), limited resources, intraspecific competition (Gurtin & Maccamy, 1974, Rorres, 1976, Poluektov et al, 1980, Swick, 1977, Webb, 1985).

1.3 Nonlinear Age-Dependent Models with Intra-Species Competition

Individuals of the same species almost always fight for food resources, territory, partner, etc. Such intra-species competition can be incorporated into a corresponding population model using integral or partial differential equations. A simple age-dependent nonlinear model of a single species population in a stationary environment with intra-species competition is described by the following *nonlinear evolutionary equation* (Moiseev, 1983; Poluektov et al, 1980):

$$\partial x / \partial \tau + \partial x / \partial t = -[d(\tau) + \int_0^T b(\tau,\xi)x(\xi,t)d\xi]x(\tau,t), \tag{5.14}$$

where the function $b(\tau,\xi)$ describes the increase in the mortality of individuals of age τ caused by individuals of age ξ (the *intra-species competition* for scarce resources). Other model functions are the same as in model (5.9). Here the death rate is a function of the population density $x(\tau,t)$ and increases when the density increases. Bifurcation and stability analysis of model (5.14) was provided in (Moiseev, 1983). A nonlinear integral model equivalent to model (5.14) was derived in (Poluektov et al, 1980) and had a complicated form.

We suggest the nonlinear biological integral model

$$X(t) = \int_{-T} m(t-\xi) e^{-d(t-\xi) - \int_{t-T}^{t} b(t-\xi, t-\theta) X(\theta) d\theta} X(\xi) d\xi \qquad (5.15)$$

that describes the same population regulatory mechanism (the death rate as a function of the population density) as the PDE model (5.14) does.

In this model, we assume that the non-linear component (intra-specific competition factor) inside the fatigue function

$$l(X(.);t) = e^{-d(t-\xi) - \int_{t-T}^{t} b(t-\xi, t-\theta) X(\theta) d\theta}$$

depends on the cumulative negative impact of all individuals at time t (i.e. the fatigue function depends implicitly on the current time t). This assumption reflects the fact that the cumulative birth-rate intensity $X(t)$ is produced at the current moment t by all individuals alive now.

Later, in Section 2 we will provide the qualitative analysis of (5.15), find its stationary solutions, verify their stability, and locate bifurcation points.

We would like to note that integral models with a simpler non-linearity

$$x(t) = \int_{-T} f(\xi, x(\xi)) d\xi$$

are often used for modelling of epidemic problems (Gurtin & Maccamy, 1974, Webb, 1985). A major interest is to find out whether a disease, once it enters a population, can persist or dies out; and how much this depends on various parameters of a model, notably; the duration of infective period. In terms of mathematics, this involves the problems of finding steady-state solutions (constant, periodic, almost periodic, etc.) and analyzing their stability (Cook & Kaplan, 1976, Leggett & Williams, 1983).

In such models the value T is interpreted as *the duration of infection*, $x(t)$ is the population of infectious individuals at time t, $f(t,x(t))$ is the *instantaneous rate of infection*, and $f(t,x(t))dt$ is the fraction of individuals infected within the period $[t, t+dt]$. It is usually assumed that $f(t,0)=0$ and $f(t,x)$ increases with respect to x (the more infectious, the more individuals can be infected). The non-linearity in such models describes effects different from the intra-specific competition. Similar problems arise in other biological applications.

1.4 Models with Delay

In most biological systems individuals born at time t do not produce descendants at once (at time t) because they need time to reach maturity, for the destination period, etc. The infection for most diseases does not spread instantly. The infective period is essential in theory of epidemics. Even the initial symptoms of a disease show up in a certain period of time. As you can see a time delay is involved and therefore it is very important to take into account not just the present state of a physical system but also its past history. Various mathematical tools can be used to deal with the time delay. Delay differential equation models and integral models are some of them. They have a richer mathematical framework (compared with ordinary differential equations) for the analysis of biosystems dynamics. Integral models of population dynamics similar to (5.15) take into account the *distributed delay* on the interval $[t-T, t)$, called a *process after-effect* or *contagion,* or *hereditary effects,* or *the time to reach maturity.* It determines the influence of the prehistory of a process at τ on the process dynamics at time t, i.e. dependence of the current time t on the prehistory distance $t-\tau$.

Other types of models may consider *the lumped (point) delay,* that is, the influence of the past instant on the current state of a process. It leads to *delay (lag) differential equations* (Gripenberg et al, 1990; Murray 1989; Poluektov et al, 1980; etc.):

$$dN(t)/dt = f(N(t), N(t-T)),$$

where $T>0$ is the *duration of the delay.* In population models, the organisms with seasonal reproduction have such behavior.

Introducing the lumped delay into the Verhulst-Pearl model (5.5) leads to the following delay differential equation:

$$dN(t)/dt = \mu N(t)\,(1 - N(t-T)\,/K). \tag{5.16}$$

Corresponding more accurate *model with distributed delay* that takes into account an average over past population is

$$\frac{dN(t)}{dt} = \mu N(t)[1-(\int_{-\infty}^{t}\beta(t-\xi)N(\xi)d\xi)/\mathrm{K}] \tag{5.17}$$

where a weighting factor $\beta(t-\xi)$ reflects the dependence of current resource availability upon the size of the population at times preceded the time t.

A time-dependent solution of delay differential equations is not uniquely determined by its initial state at a given moment. The solution profile on an interval with the length equal to the maximal delay T has to be given as an initial condition. Therefore even a single delay differential equation turns to be an infinite-dimensional problem.

Steady-state solutions of delay differential equations are the same as those of the corresponding system of ordinary differential equations where all delays are set up to zero. Bifurcation analysis of steady-state solutions can also be provided from the corresponding ordinary differential equation system. The stability analysis is different. Comparing the delay differential equation (5.16) with its non-delay analogue, we can find that the stationary solution of (5.16) $N(t) = K$ is stable if $0< \mu T< \lambda$ and unstable for $\mu T>\lambda$, where λ is a bifurcation value. Moreover, the solution oscillates around the environment capacity value K and reveals stable limit cycle behavior for some values $\mu T>\lambda$. Non-delay differential equation models like (5.5) and (5.2) cannot exhibit a periodic behavior. Therefore, adding a delay to a differential equation decreases its stability and allows for oscillatory regimes. The latter feature motivates the use of delay models to study the dynamics of populations with periodic behavior.

1.5 Difference Models

Another important type of models with the lumped delay is represented by *difference equations*. There is a wide class of biological problems reducible to such equations. They have been studied in depth by many scientists and applied to different areas. The difference equations may have good reasoning to be chosen for a real-life process description or may be obtained by discretization of differential or integral equations. In Section 2.3 we provide two examples when the integral model (5.15) is transformed to difference equation models in a special case.

A simple non-linear difference equation of the first order has the following form:

$$X(t+1) = f(X(t)), \quad t = 0,1,2,\dots . \tag{5.18}$$

or another form convenient for showing the existence of stationary solutions:

$$X(t+1) = X(t) \cdot F(X(t)), \tag{5.19}$$

where

$X(t)$ is the size of a population at time t;
$f(X(t))$ and $F(X(t))$ are functions of $X(t)$, usually nonlinear.

The case $F(X(t)) = 1+\mu t$ produces a difference approximation of the Malthus model (5.2); the case $F(X(t)) = r/(1+X(t)/K)$ refers to the Verhulst-Pearl model (5.5) discussed in Section 1.1.

Setting $X(t+1) = X(t) = X^*$ in (5.19) and solving the resulting equation

$$X^* \cdot (F(X^*) - 1) = 0$$

for X^*, one can find that stationary solutions of (5.19) are $X^* = 0$ and X^* from $F(X^*) = 1$. A graphical procedure for solving equation (5.19) (so-called *cobwebbing*) consists of finding the intersections of the straight line $X(t+1) = X(t)$ and the curve $X(t+1) = X(t) \cdot F(X(t))$ (starting with an initial state $X(0) = X_0$).

To investigate the linear stability of stationary solutions (Murray, 1989) we have to examine the behavior of the first derivative $f'(X^*)$ at a stationary point X^*, called the *eigenvector* of the system at the stationary solution X^*. If

$$\left| \, dN(t)/dt \,\big|_{X(t)=X^*} \right| = \left| f'(X^*) \right| < 1,$$

then the stationary solution is linearly stable or *attracting*. It means that a solution $X(t)$ obtained by any small perturbation from a stationary solution X^* tends toward X^* in the long run or $X(t) \to X^*$ as $t \to \propto$. If the solution $X(t)$ instead of approaching the stationary solution X^* goes far X^*, then the stationary solution X^* is unstable or *repelling*. In this case $f'(X^*) < -1$. The case of a neutrally stable stationary solution X^* occurs when $|f'(X^*)| = 1$. The values $\lambda = \pm 1$ are bifurcation values since there is a qualitative change in the behavior of a solution of (5.18) when λ crosses 1 or -1. Moreover, the case of $\lambda = 1$ represents a tangent bifurcation; $\lambda = -1$ the case of a pitchfork bifurcation.

The difference models possess really remarkable bifurcations. Their bifurcations have been thoroughly studied in many publications during the last two decades. Some specific patterns of the dynamics have been discovered for special types of such models (first-order, second-order, etc.). In particular, already in the case of simple quadratic non-linearity

$$X(t+1) = AX(1-X). \tag{5.20}$$

Model (5.20) is called *discrete logistic* equation or *May* model. It has been exposed in many publications (May, 1976; Casti, 1989; Poluektov et al, 1980 etc.) that when the parameter A increases from 0 to 4, the solution of (5.20) is changed vastly: from a single zero stationary state through a stable non-zero stationary state to oscillations (stable cycles) with increasing periods, instability, and extremely complicated quasi-chaotic behavior (indistinguishable from random one when time is large). Such dynamics appears to be natural for similar types of the non-linearity $f(X)$.

The difference model

$$X(t+1) = Ae^{-BX(t)}\, X(t) \tag{5.21}$$

was introduced by Ricker in 1954. Despite its solid age, the *Ricker model* is still popular in biological literature (Edelstein-Keshet, 1988; Fulford et al, 1997; Hoppensteadt & Peskin, 1992). It has been used in numerous applications and gives a fair correspondence to experimental data for various populations (in particular, fish).

A stationary solution (trajectory) $f(X)=X$ change equation (5.21) to

$$X(Ae^{-BX}-1)=0.$$

The last equation has two stationary solutions: $X=0$ and $X=\ln A/B$. Let us provide the initial bifurcation analysis of the Ricker model (5.21) (Sharkovsky et al, 1994). The investigation technique for difference equations employs properties of the one-dimensional map associated with the difference equation. The behavior of a periodic solution (a cycle) $\Lambda=\{\beta_1, \beta_2,\ldots, \beta_n,\beta_{1+i}, \beta_{2+i},\ldots, \beta_{n+i}\,,\ i=0,\ldots,\infty\}$ of the difference equation (5.18) is determined by its "spectrum multiplier" or eigenfunction $\mu(\Lambda)= f'(\beta_1)f'(\beta_2)\ldots f'(\beta_n)$. If $|\mu(\Lambda)|<1$, then solution Λ is attracting, if $|\mu(\Lambda)|>1$, then solution Λ is repelling. In our case $f'(X) = A\,e^{-BX}(1-BX)$, and equation (5.21) has the following qualitative dynamics for various ranges of the parameter A:

(1) $0 < A < 1$, there is a single fixed point $\beta_0 = 0$ and it is attracting,

because $|\mu(A)| = |f'(0)| = C < 1$.

(2) $1 < A < e^2$, the fixed point $\beta_0 = 0$ becomes repelling ($f'(0) = A > 1$); a new positive fixed point $\beta_1 = \ln A/B$ appears. The multiplier $\mu(\beta_1) = f'(\beta_1) = 1 - \ln A < 0$, which means that the trajectory oscillates around the value β_1 and tends to it (since $|\mu(\beta_1)|<1$).

(3) $e^2 < A < e^2 + o(e^2)$, the fixed point $\beta_1 = \ln A/B$ becomes repelling ($\mu(\beta_1) < -1$) and gives the birth to the periodic solution (cycle) $\{\beta_1, \beta_2\}$ with period 2.

(4) $\ln A > 2$, there are two solutions $\beta_{2,3} = \ln A/B + \varepsilon_{2,3}/B + o((\ln A - 2)/B)$ different from β_0 and β_1; where $\varepsilon_{2,3}$ are defined from the quadratic equation $\varepsilon^2(3 - \ln A)/(6 - 3\ln A) - \varepsilon^2 + 1 = 0$. The analysis shows that the multiplier $|\mu(\beta_1, \beta_2)| = |f'(\beta_1) f'(\beta_2)| < 1$ at $\ln A - 2 << 1$, so the cycle is attracting.

With the further growth of the parameter A, the cycle becomes repelling and cycles with increasing periods appear with the further transition to a chaotic motion. The parameter A has the bifurcation values A_0 and A_1 such that the equation (5.18) has the stable fixed point $X^0=0$ for $0<A<A_0$ and a stable positive fixed point $X^*>0$ for $A_0 \le A < A_1$. At $A = A_1$ the stability of X^* breaks down and a stable cycle of period 2 appears for $A_1 \le A < A_2$. Again at $A = A_2$ the stability of this cycle breaks down and a stable cycle of the period 4 appears. This process continues indefinitely with the cycles of period 2^k arising at $A = A_k$ and the sequence $\{A_k\}$ converges to a value A_∞. The set of parameters $\{A_k\}$ are naturally different for various functions $f(X)$ but some universal proportions among them have been also discovered. At $A = A_\infty$ the cycles with all periods 2^k, $k=1,2,\ldots$, coexist and are repelling, the equation has no attracting cycles and each of its trajectory are almost periodic. At $A > A_\infty$ new cycles appear with the periods different from 2^k, and, when the parameter A increases further, at some point A_c the equation has cycles with every integer period. At $A > A_c$ there are trajectories with arbitrarily great periods and an uncountable number of aperiodic trajectories, which coexist and never intersect. There is a set of values of the parameter A, $A > A_c$, when the equation (5.18) does not have any stable cycle but so-called "*strange attractors*" that consist of a finite number of intervals and the trajectory "randomly" jumps among these intervals.

Such behavior appears to be natural for the first-order difference equation (5.18) in general case. Namely, the below qualitative picture is valid for almost all *unimodal* functions $Af(X)$ (Casti, 1989; Sharkovsky et al, 1994).

1.6 Spatial Diffusion Models of Population Dynamics

Diffusion theory is ideal for describing and investigating quite a large number of biological situations, ranging from neural activity and spread of diseases to patterns of movement of biological population, since the dynamics of the most biosystems exhibit spatial distributions. In addition to the description of effects of a spatial variation, diffusion models lead insight into various fundamental population processes like dispersal, ecological invasions, the effect of habitat geometry, and dispersal-mediated coexistence, providing a natural way for investigating and understanding the interaction among habitat/environment, individual behavior, and population dynamics. They assist in making valid predictions about population density and diversity. Diffusion equations are also beneficial because they allow incorporating both temporal and spatial processes simultaneously into equations governing population dynamics (Murray, 1989; Banks et al, 1988).

Among other factors, mechanisms of dispersal include making small random moves due to environment changes or external attractions. Study of such *random walk models* helps to identify and display features and ways of changes of a spatial process and make necessary conditions and forecasts. Moreover, random walk models are easy to derive and imagine.

The diffusion equation has been derived in Section 2.3 using conservation laws applied to general physical continuum media. In this section we show the interrelation between Markov random and diffusion models by converting a random-walk equation into a diffusion equation.

1.6.1 Random Walk Models

One of the general mathematical tools used in many applications is a *Markov process* when the current position of a system depends only on the immediately preceding one and not on any other past positions. Introduced by the Russian mathematician Andrei Andreyevich Markov (1856-1922), Markov processes have captured the interest of scientists and have been developed fast in the last decades. It has a wide range of applications from the study of the evolution of genetic traits, statistical models of protein and DNA sequences to spread of infection, migration of animals and speech recognition. In this section, we will discuss *Markov random walk models.*

Various particles from small cells, germs, bacteria, viruses to large animals move around and spread out in a random way. Random phenomena such as interaction between objects, food or habitat preferences, environment etc. can influence their motion and change their behavior. These various random phenomena can be described by adding probabilities of all possible changes to the model. At the beginning, we assume that the changes and their probabilities depend only on the previous state of a system and take a one-dimension movement. Then, an object moves randomly only right and left (back and forth) or performs *random walk* that is a particular Markov process.

To illustrate this process, let us consider an example with trees infested with an insect population. If trees are in a line (along a river, lake, road, etc), then insects move only to the right and left tree. For simplicity, we assume that the size of the insect population is the same and its motion is unbiased. We also suppose that it is possible to estimate the proportion of the insects $s(i,t)$ in tree i at time t, $\Sigma_t\, s(i,t) = 1$, and the probability p of moving one insect to the nearest tree on the right or left. Thus $(1\text{-}2p)$ gives the probability that an insect stays in the tree. Then at time $t+1$, tree i has its own insects plus insects from its left-side tree i-1 and right-side tree i+1, or the proportion of insects that tree i has at time t+1 is:

$$s(i, t+1) = p\, s(i-1, t) + (1-2p)\, s(i, t) + p\, s(i+1, t) \tag{5.22}$$

or in matrix notation:

$$\mathbf{s}_{n+1} = \mathbf{s}_n \boldsymbol{P}, \tag{5.23}$$

$$P = \begin{bmatrix} \cdots & \cdots & \cdots & \cdots & \cdots & \cdots & \cdots \\ \cdots & p & 1-2p & p & \cdots & 0 & \cdots \\ \cdots & 0 & p & 1-2p & \cdots & 0 & \cdots \\ \cdots & 0 & 0 & p & \cdots & 0 & \cdots \\ \cdots & \cdots & \cdots & \cdots & \cdots & \cdots & \cdots \\ \cdots & 0 & 0 & 0 & \cdots & p & \cdots \\ \cdots & \cdots & \cdots & \cdots & \cdots & \cdots & \cdots \end{bmatrix} \tag{5.24}$$

Here

$\mathbf{s}_n = (\ldots, s(-i-1,t), s(-i,t), s(-i+1,t), \ldots, s(0,t), \ldots, s(i-1,t), s(i,t), s(i+1,t), \ldots)$ describes the probability distribution of the insects among the trees;

the tree 0 is a randomly picked initial tree,
the tree i represents the i-th tree to the right of the initial tree,
the tree $-i$ represents the i-th tree to the left.

The matrix $\boldsymbol{P}$ is called the one-step *transition matrix* and shows the probability of moving from state i to state j in one time period. It plays a key role in the theory and application of Markov processes and has incredible properties. For example, the sum of all entries in a row of $\boldsymbol{P}$ is always 1; matrix $\boldsymbol{P}^k$ is called the k-step transition matrix and defines the probability of being in the j-state k time periods after being in state i. If there is $l > 1$ such as the matrix $\boldsymbol{P}^l$ has only nonzero entries, then the Markov process is regular and a path that connects any two states can always be found. If such a situation occurs, then a system has a stable steady solution.

Model (5.22) or (5.23) - (5.24) gives a qualitative description of particles (in our case, the infestation) with the unlimited number of elements (trees) in a population. If there are a finite number of elements, say, N elements to the left and N elements to the right of the initial one, then different situations occurred at the ends of the line can be described by different boundary conditions corresponding to each case. We emphasize our attention on the most common and natural situations:

Case 1. Periodic boundary conditions. Elements of a population form a closed curve (trees surround a lake or a square). In this case, the elements N and $-N$ are neighbors and replacing

$$s(N+1,t) = s(-N,t) \quad \text{and} \quad s(-N-1,t) = s(N,t)$$

we get *periodic boundary conditions*. The transition probability matrix $\boldsymbol{P}$ in (5.23) turns to be

$$P = \begin{bmatrix} 1-2p & p & 0 & \dots & 0 & 0 & p \\ p & 1-2p & p & \dots & 0 & 0 & 0 \\ 0 & p & 1-2p & \dots & 0 & 0 & 0 \\ 0 & 0 & p & \dots & 0 & 0 & 0 \\ \dots & \dots & \dots & \dots & \dots & \dots & \dots \\ 0 & 0 & 0 & \dots & p & 1-2p & p \\ p & 0 & 0 & \dots & 0 & p & 1-2p \end{bmatrix} \qquad (5.25)$$

Model (5.23) - (5.25) is called a *random walk Markov chain on a circle* and matrix $\boldsymbol{P}$ is called a doubly stochastic matrix since not only each row of the matrix adds up to 1 but also each column.

If elements of a population are in a line (trees are along a river or a road), then various situations are possible. We discuss just two of them.

Case 2*a*. Absorbing boundary conditions. An insect reaches end elements and disappears, that is there are two end elements called *absorbing states* that absorb all insects (for example, an insect-killer is placed). *Absorbing boundary conditions* reflect this case. The transition matrix $\boldsymbol{P}$ will take the form:

$$P = \begin{bmatrix} 1 & 0 & 0 & \dots & 0 & 0 & 0 \\ p & 1-2p & p & \dots & 0 & 0 & 0 \\ 0 & p & 1-2p & \dots & 0 & 0 & 0 \\ 0 & 0 & p & \dots & 0 & 0 & 0 \\ \dots & \dots & \dots & \dots & \dots & \dots & \dots \\ 0 & 0 & 0 & \dots & p & 1-2p & p \\ 0 & 0 & 0 & \dots & 0 & 0 & 1 \end{bmatrix}$$

where the first and the last rows represent the absorbing states. There is one 1 at the diagonal position and all other elements are zeroes in those rows. It means that it is impossible to go to any other state after reaching the absorbing states. On the other hand, there is a path for each insect to reach an absorbing state from any other state for a finite number of steps. The process is not regular because the matrix P is not regular. Its submatrix Q that consists of nonabsorbing states is important in the theory of Markov processes.

Case 2*b*. Reflecting boundary conditions. An insect reaches end elements and stays there or goes back, that is end trees reflect insects (a screen is placed next to each of the end trees). The resulting *reflecting boundary conditions*

$$s(N+1,t) = s(N-1,t) \quad \text{and} \quad s(-N-1,t) = s(-N+1,t)$$

change the transition matrix $\boldsymbol{P}$ to

$$P = \begin{bmatrix} 1-p & p & 0 & 0 & \dots & 0 & 0 \\ p & 1-2p & p & 0 & \dots & 0 & 0 \\ 0 & p & 1-2p & p & \dots & 0 & 0 \\ 0 & 0 & p & 1-2p & \dots & 0 & 0 \\ \dots & \dots & \dots & \dots & \dots & \dots & \dots \\ 0 & 0 & 0 & 0 & \dots & 1-2p & p \\ 0 & 0 & 0 & 0 & \dots & p & 1-p \end{bmatrix}$$

The theory of Markov processes has been developed pretty well. Various techniques are based on remarkable properties of the transition matrix $\boldsymbol{P}$. We just want to mention a few of them: k-step transition matrix $\boldsymbol{P}^k$ shows the behavior of a system in k time periods; the steady-state probability vector π from $\pi = \boldsymbol{P'}\pi$ gives the proportion of insects in each element (tree) in the long run in cases **1** and **2*b***; the mean time of absorption $\mu = (\boldsymbol{I}\text{-}\boldsymbol{Q})^{-1}\cdot\mathbf{1}$, (where $\boldsymbol{I}$ is the identity matrix, **1** is a vector made up from 1) estimates the amount of time it takes each insect to reach an absorbing state or die out in case **2*a***; and so on.

Relatively easy simulation of Markov processes helps to study the effect of small fluctuation in a system and determine how sensitive or robust the system is to the various changes. Also some one variable non-Markov processes with stochastic, deviation or noise in a physical system can be transformed to a multi-variable Markov process.

1.6.2 Diffusion Models

Although models involving random walks are easy to understand, derive, investigate and provide computer simulation it is still difficult to get their analytic solution. That is why converting random-walk models into differential equations that can be solved is useful. In the following section we show the interrelation between Markov and differential models.

Suppose that the number N of elements of a population is large and the elements are close to each other, that is the distance $\delta x = 1/N$ between two elements $x_i = i/N$ and x_{i+1} is small; the variable t stands for the time scale; the sampling time δt is short; $u(x, t)$ describes the concentration of particles and is the analogue of $s(i,t)$ from Section 1.6.1. Model (5.22) in these notations takes the form:

$$u(x,t+\delta t) = p\ u(x-\delta x,t) + (1\text{-}2p)\ u(x,t) + p\ u(x+\delta x,t)$$

and after providing simple mathematics becomes

$$\frac{u(x,t+\delta t)-u(x,t)}{\delta t} = D\frac{u(x-\delta x,t)-2u(x,t)+u(x+\delta x,t)}{\delta x^2}$$

where a constant $D=p\delta x^2/\delta t$ is called the *diffusion coefficient* or *diffusivity* and measures dispersal rate with units distance^2 /time.

Letting δt and δx approach 0 in the last equation, we obtain the one-dimensional diffusion equation

$$\frac{\partial u}{\partial t} = D\frac{\partial^2 u}{\partial x^2} \tag{5.26}$$

Model (5.26) arises in many engineering and life sciences applications (chemical diffusion, heat flow, molecules movement, neural activity, etc.) where objects exhibit Brownian motion. It measures the density of a particle as a function of time and position. The solution to (5.26)

$$u(x,t) = \frac{e^{-\frac{x^2}{4Dt}}}{\sqrt{4\pi Dt}} \tag{5.27}$$

is the probability density function of the normal distribution with the mean 0 and the variance $2Dt$ (the standard deviation $\sqrt{2Dt}$). Thus, the diffusion coefficient D proportional to the probability (proportion) p is related to the variance of the random walk. The graph of the solution (5.27) is a well-known bell-shaped normal distribution curve. Since the solution is described by a normal curve, then there is a small finite probability that an individual travels an arbitrary long distance in arbitrary short time. To calculate the probability that x is in the interval $[a,b]$, it is necessary to evaluate:

$$\int_a^b \frac{e^{-\frac{x^2}{4Dt}}}{\sqrt{4\pi Dt}}\,dx$$

or to use the standard normal distribution table after prior normalization (5.27):

$$Z = x/(\sqrt{2}Dt).$$

The diffusion equation (5.26) is the continuous analogue of the random walk model (5.23)-(5.24) in unlimited case. A diffusion equation can be associated with each Markov chain. Although simple closed-form solutions are not known in most cases with boundary solutions, investigating diffusion equation (5.26) with boundary conditions we get results similar to random-walk models - in the reflecting and periodic cases. The solution approaches a constant as $t \to \infty$; in the absorbing case, the solution approaches zero as $t \to \infty$.

Similar calculations show that in a two-dimension case, assuming that $\delta x^2 / \delta t \to D$; $\delta y^2 / \delta t \to D$, we get the equation:

$$\frac{\partial u}{\partial t} = D(\frac{\partial^2 u}{\partial x^2} + \frac{\partial^2 u}{\partial y^2}) \tag{5.28}$$

Using model (5.28), it is possible to investigate different processes, from examining a process spread of a nutrient drop placed in the center of a circular petri dish to prediction of the location of a single individual at some particular distance from a starting point. The diffusion equation has been used to successful description of the movement of animals in mark-recapture study in numerous experiments, that is, animals realized at a certain point disperse via diffusion in a two-dimensional environment.

Models (5.26) and (5.28) work best when the environment is homogeneous and the individuals in the populations have similar random Brownian motion. Nonrandom or biased random movements influence the outcome and require different models. Models (5.26) and (5.28) can be easily adjusted and modified in response to environment changes. Let us show just a few examples (Holmes et al, 1994).

When animals (organisms) are influenced by stimuli or carried by the wind or water currents, with drift velocities w_x and w_y, then (5.28) changes to *advection-diffusion equation* (see also Section 3.4 in Chapter 2):

$$\frac{\partial u}{\partial t} = D(\frac{\partial^2 u}{\partial x^2} + \frac{\partial^2 u}{\partial y^2}) - w_x \frac{\partial u}{\partial x} - w_y \frac{\partial u}{\partial y}$$

If animals (organisms) are attracted to one another ($k>0$) or repelled from one another ($k<0$), then a *nonlinear diffusion equation* is obtained:

$$\frac{\partial u}{\partial t} = D\frac{\partial^2 u}{\partial x^2} + \frac{\partial}{\partial x}(ku\frac{\partial u}{\partial x})$$

If animals prefer an existing well-known path and tend toward it, then using mathematical language, they provide a correlated random walk with finite animal velocity v:

$$\frac{\partial u}{\partial t} = \frac{v^2}{2\rho}(\frac{\partial^2 u}{\partial x^2} + \frac{\partial^2 u}{\partial y^2}) - \frac{1}{2\rho}\frac{\partial^2 u}{\partial t^2}$$

where $1/2\rho$ is a correlation coefficient between the directions of travel.

Scientists are paying increased attention to spatial processes in a wide variety of practical contexts and have studied them well. Diffusion equations have been used to model such a process for a large variety of biological systems. They work well when spatial homogeneity is assumed and, as numerical studies have shown, in cases of spatial inhomogeneity. The robustness of a diffusion equation solution provides a forceful guidance to researchers and increases the number of applications.

2. BIFURCATION ANALYSIS FOR NONLINEAR INTEGRAL MODELS

Bifurcation analysis of nonlinear ecological models is based on investigation of their *stationary solutions* and the *bifurcation values* of parameters at which new stationary solutions appear or the qualitative behavior of a solution changes. We have chosen a less studied non-linear integral model to provide qualitative analysis and illustrate basic ideas and features of the bifurcation analysis.

As the basic result of this section we are going to prove that the non-linear integral model (5.15) has a bifurcation at $\lambda=\lambda^*$ and show that there are three branches $X(\lambda)$ of the stationary solutions of the model in some neighborhood of the value λ^*:

- one branch at $0<\lambda<\lambda^*$ - stable trivial solution X^0_S,
- two branches at $\lambda>\lambda^*$ - unstable zero solution $X^0_S=0$

  stable positive solution $X^*_S>0$.

Moreover, the positive stationary solution $X^*_S>0$ are stable for $\lambda^*<\lambda<\lambda^{**}$ and unstable for $\lambda>\lambda^{**}$, where λ^{**} is the second critical value of the parameter λ. λ^*, λ^{**} will become clear below.

2.1 Stationary Solutions

To find stationary solutions let us introduce a parameter into non-linear integral model (5.15). All basic functions such as mortality and death rates etc. depend on parameters that follow from various applications and arise naturally. We suppose that the fertility rate

$$m(\tau) = \lambda\ \overline{m}(\tau)$$

where λ is some parameter that determines the intensity of fertility but does not change its structure. We may consider the parameter λ to be directly proportional to food amount or other characteristics essential to a population.

Then after introduction λ and substitutions $\tau = t-\xi$; $\xi = t-\theta$ Model (5.15) is rewritten as:

$$X(t) = \lambda \int_0^T \overline{m}(\tau) e^{-d(\tau)-\int_0^T b(\tau,\xi)X(t-\xi)d\xi} X(t-\tau)d\tau. \qquad (5.29)$$

The stationary states of models of type (5.29) correspond to the *constant total fertility* $X(t)\equiv X_S=\text{const}\geq 0$. For their study, it is sufficient to substitute X_S into the original model and solve the obtained nonlinear equation. The same technique is widely used for differential and difference models.

Substituting $X(t)\equiv X_S$ =const into (5.29) we obtain the following nonlinear equation for the stationary solutions X_S :

$$X_S\{1 - \lambda \int_0^T \overline{m}(\tau) e^{-d(\tau)-X_S\int_0^T b(\tau,\xi)d\xi} d\tau\} = 0. \qquad (5.30)$$

It follows from (5.30) that there is the trivial stationary state $X^0_S=0$ for any $\lambda>0$. Moreover, it is the only stationary state for small values of λ. The second stationary state occurs as λ increases up to some value $\lambda=\lambda^*$ that is determined by the following equality

$$R(\lambda^*) = 1, \tag{5.31}$$

where

$$R(\lambda) = \lambda \int_0^T \overline{m}(\tau) e^{-d(\tau)} d\tau. \tag{5.32}$$

The value $R(\lambda)$ is called the *reproduction number* or *biological potential of a population* and describes the average number of descendants of one individual during its lifetime. For stable natural populations $R\approx 1$.

Thus the value $\lambda=\lambda^*$ turns to be a *bifurcation value*. Namely, for $\lambda>\lambda^*$ a new nontrivial stationary state X^*_S appears. In other words, the point $(\lambda^*,0)$ of the plane (λ,X) is the *ramification point* of the nonlinear equation (5.30) and the following result has been obtained.

There are three branches $X(\lambda)$ of stationary solutions of the model (5.29) in some neighborhood of the value λ^*:

- one branch at $0<\lambda<\lambda^*$ - trivial solution X^0_S,
- two branches at $\lambda>\lambda^*$ - trivial solution $X^0_S=0$
  nontrivial positive solution $X^*_S>0$.

2.2 Stability Analysis

In the previous section, we found all stationary solutions and a bifurcation value of the parameter λ. Now we are going to investigate the stability of stationary solutions and verify if there is only one bifurcation value of the parameter λ^*.

Stability analysis is provided in assumption that the given functions $m(\tau)$, $l(\tau)$, $b(\tau, t)$ in equation (5.29) are continuous for $(\tau,t)\in[0,T]\otimes[0,T]$ and solutions $X(t)$ are continuous on the interval $[0,T]$. Naturally, the initial function $X_0(\xi)=\phi(-\xi)$, $\xi\in[-T,0]$, $T>0$ is also continuous.

The linear stability analysis of stationary solutions is provided in several steps:

(1) obtaining the linearization form of non-linear integral model (5.29);
(2) investigating the stability of the trivial stationary solution and finally the nontrivial one.

To find the linear equation in variations for the non-linear integral model (5.29), we give small increments $\delta X(t)$ and $\delta X_0(t)$ to the sought-for variable $X(t)$ and the given initial function $X_0(\tau)$, i.e. substitute the function $X(t)+\delta X(t)$, $t\in[0,T)$ and $X_0(\tau)+\delta X_0(\tau)$, $\tau\in[-T,0)$ into equation (5.29):

$$X(t)+\delta X(t)=$$
$$=\lambda\int_0^T \overline{m}(\tau)e^{-d(\tau)-\int_0^T b(\tau,\xi)(X(t-\xi)+\delta X(t-\xi))d\xi}(X(t-\tau)+\delta X(t-\tau)d\tau.$$

Substituting (5.29) for $X(t)$ we obtain the following equation for $\delta X(t)$:

$$\delta X(t)=$$
$$=\lambda\int_0^T \overline{m}(\tau)e^{-d(\tau)-\int_0^T b(\tau,\xi)X(t-\xi)d\xi}\times \tag{5.33}$$
$$\times(e^{-\int_0^t b(\tau,\xi)\delta X(t-\xi)d\xi-\int_t^T b(\tau,\xi)\delta X_0(t-\xi)d\xi}-1)X(t-\tau)d\tau+$$
$$+\lambda\int_0^t \overline{m}(\tau)e^{-d(\tau)-\int_0^T b(\tau,\xi)X(t-\xi)d\xi-\int_0^T b(\tau,\xi)\delta X(t-\xi)d\xi}\delta X(t-\tau)d\tau+$$
$$+\lambda\int_t^T \overline{m}(\tau)e^{-d(\tau)-\int_0^T b(\tau,\xi)X(t-\xi)d\xi-\int_0^T b(\tau,\xi)\delta X(t-\xi)d\xi}\delta X_0(t-\tau)d\tau.$$

Since $|\delta X(t)|<<1$ all terms involving powers higher than $\delta^2 X(t)$ are negligible in comparison with the terms involving $\delta X(t)$. Thus omitting all high powers of $\delta X(.)$ from (5.33) and using Taylor extension for e^x we get:

$$\delta X(t)=$$
$$-\lambda\int_0^T \overline{m}(\tau)e^{-d(\tau)-\int_0^T b(\tau,\xi)X(t-\xi)d\xi}X(t-\tau)\int_0^t b(\tau,\xi)\delta X(t-\xi)d\xi d\tau\ \_$$
$$-\lambda\int_0^T \overline{m}(\tau)e^{-d(\tau)-\int_0^T b(\tau,\xi)X(t-\xi)d\xi}X(t-\tau)\int_t^T b(\tau,\xi)\delta X_0(t-\xi)d\xi d\tau+$$
$$+\lambda\int_0^t \overline{m}(\tau)e^{-d(\tau)-\int_0^T b(\tau,\xi)X(t-\xi)d\xi}\delta X(t-\tau)d\tau+$$
$$+\lambda\int_t^T \overline{m}(\tau)e^{-d(\tau)-\int_0^T b(\tau,\xi)X(t-\xi)d\xi}\delta X_0(t-\tau)d\tau+o(|\delta X|).$$

Changing the order of integration in the first two integrals leads to the *linear equation in variations* for the non-linear integral model (5.29):

$$\delta X(t) \approx \lambda \int_0 K(\tau,t)\delta X(t-\tau)d\tau + \lambda \int^T K(\tau,t)\delta X_0(t-\tau)d\tau, \quad (5.34)$$

where

$$K(\tau,t) = \overline{m}(\tau)e^{-d(\tau)-\int_0^T b(\tau,\xi)X(t-\xi)d\xi} -$$
$$\int_0^T \overline{m}(u)e^{-d(u)-\int_0^T b(u,\xi)X(t-\xi)d\xi} b(u,\tau)X(t-u)du \quad (5.35)$$

The first step of the stability analysis has been completed.

In conclusion, we would like to notice that equation (5.34) is a linear Volterra integral equation of the second kind with respect to $\delta X(t)$, $t \in [0,T)$. For any solution of a linear Volterra integral equation,

$$x(t) = \int_0 K(\tau,t)x(\tau)d\tau + f(t)$$

the Gronwall-Bellman inequality (see Appendix, Section 2.2.4) is true:

$$| x(t) | \le e^{\|K\|(t-t_0)} \|f\|. \quad (5.36)$$

This result will be used later. Also in future discussions the standard norm of a continuous function $x(t)$ is used:

$$\|x\| = \{\max | x(t) |, t \in D_x\}, \quad (5.37)$$

where D_x is the domain of the function x; i.e., $D_{X0} = [-T,0]$, $D_m = [0, T]$.

Now, we start the second step of the stability analysis. At the beginning, let us investigate the stability of the trivial stationary solution $X^0_S = 0$. As it has been shown above, λ^* is a bifurcation point where a solution changes its qualitative behavior. Thus it is wise to divide the analysis of stability of the trivial solution into two parts, namely $0<\lambda<\lambda^*$ and $\lambda>\lambda^*$. The proof of stability in both cases is based on using the linear equation in variations (5.34),(5.35) where the function $K(\tau,t)$ related to $X^0{}_S=0$ turns to be:

$$K(\tau,t)=K_0(\tau)=\overline{m}(\tau)e^{-d(\tau)} \tag{5.38}$$

The trivial solution.

Case 1: $0<\lambda<\lambda^*$. Let us find the upper bound of $\delta X(t)$ as t increases and a small perturbation $\delta X_0\,(t)$ is given to the initial value. Applying (5.36) and (5.37) to the solution of the equation in variations (5.34) and using (5.33) and (5.38), we obtain that

$$|\,\delta X(t)\,|\le e^{\lambda\|K\|t}\lambda\left\|\int_0^T K(\tau,t)\delta X_0(t-\tau)d\tau\right\|<$$

$$<e^{\lambda\|\overline{m}\|t}\left\|\delta X_0\right\|\lambda\int_0^T\overline{m}(\tau)e^{-d(\tau)}d\tau=e^{\lambda\|\overline{m}\|t}R(\lambda)\left\|\delta X_0\right\|.$$

By Formula (5.32) $R(\lambda^*)=1$ and $R(\lambda)=\lambda/\lambda^*<1$, then

$$|\,\delta X(t)\,|<\left\|\delta X_0\right\| \tag{5.39}$$

$$\text{for}\quad t\in[0,t_1],t_1=\ln(\lambda^*/\lambda)/(\lambda\|\overline{m}\|)$$

Suppose that we have solved the equation (5.34) for $t\in[0,\ t_1]$ and supplemented the prehistory interval $[-T,\ 0]$ to $[-T,\ t_1]$. It is easy to see that the inequality (5.39) holds true for $t\in[t_1,\ t_2]$ where $t_2>t_1$ and so on. Using this line of reasoning, we can distribute this property for any limited interval. In other words small perturbations about the the trivial stationary solution $X^*{}_S=0$ lead to decreasing $\delta X(t)$ as t increases or they are not essential in the long run. Therefore, the trivial stationary solution $X^*{}_S=0$ is stable as $0<\lambda<\lambda^*$.

Case 2: $\lambda>\lambda^*$. Let us invistigate the behavior of $\delta X(t)$ as t increases after assigning a small perturbation $\delta X_0\,(t)$ to the initial value:

$$\delta X_0(\tau) = \begin{cases} \varepsilon > 0, & \tau \in [-T, -\Delta], \\ -\varepsilon\tau/\Delta + \delta X_0(0)(\Delta+\tau)/\Delta, & \tau \in [-\Delta, 0], \end{cases} \tag{5.40}$$

where $\varepsilon > 0$ represents increments of the initial value, $\varepsilon \ll 1$, $\Delta > 0$, $\Delta \ll T$. The second equation is introduced for the sake of continuity since we can use only admissible continuous variations $\delta X_0(\tau), \tau \in [-T, 0]$, i.e. those that satisfy the equation (5.34) at $t = 0$.

The value $\delta X_0(0)$ is determined from the equation (5.34) at $t = 0$, and considering (5.38) and (5.40):

$$\begin{aligned} \delta X(0) &= \\ &= \lambda \int_0^0 \overline{m}(\tau) e^{-d(\tau)} \delta X(-\tau) d\tau + \lambda \int_0^T \overline{m}(\tau) e^{-d(\tau)} \delta X_0(-\tau) d\tau = \\ &= -\lambda \int_{-T}^{-\Delta} \overline{m}(-\tau) e^{-d(-\tau)} \varepsilon d\tau + \lambda \int_{-\Delta}^0 \overline{m}(-\tau) e^{-d(-\tau)} \frac{\varepsilon\tau}{\Delta} d\tau - \\ &- \lambda \delta X_0(0) \int_{-\Delta}^0 \overline{m}(-\tau) e^{-d(-\tau)} d\tau - \lambda \delta X_0(0) \int_{-\Delta}^0 \overline{m}(-\tau) e^{-d(-\tau)} \frac{\tau}{\Delta} d\tau \end{aligned}$$

Since $\delta X(0) = \delta X_0(0)$, we obtain

$$\begin{aligned} &\delta X(0) \cdot [1 + \lambda \int_{-\Delta}^0 \overline{m}(-\tau) e^{-d(-\tau)} d\tau + \frac{\lambda}{\Delta} \int_{-\Delta}^0 \overline{m}(-\tau) e^{-d(-\tau)} \tau d\tau\,] = \\ &= \lambda\varepsilon[- \int_{-T}^{-\Delta} \overline{m}(-\tau) e^{-d(-\tau)} d\tau + \frac{1}{\Delta} \int_{-\Delta}^0 \overline{m}(-\tau) e^{-d(-\tau)} \tau d\tau \end{aligned}$$

From the last equation we can conclude that $\delta X(0) = \delta X_0(0) > \varepsilon$ at $\lambda > \lambda^*$.
Hence $\delta X_0(\tau) \geq \varepsilon$ for $\tau \in [-T, 0]$. Moreover

$$\begin{aligned} \delta X_0(0) &= \lambda \int_{-\Delta}^0 K_0(-\tau) \delta X_0(\tau) d\tau + \varepsilon\lambda \int_{-T}^{-\Delta} K_0(-\tau) d\tau > \\ &> \varepsilon\lambda \int_{-T}^0 K_0(-\tau) d\tau = \varepsilon\lambda \int_{-T}^0 \overline{m}(\tau) e^{-d(-\tau)} d\tau = \varepsilon\lambda/\lambda^*. \end{aligned} \tag{5.41}$$

Substituting the function $\delta X_0(\tau)$, $\tau \in [-T, 0]$, into the linear equation (5.34) in variations and introducing $u = t - \tau$, we obtain the following inequality:

$$\delta X(t) = \lambda \int_0 K_0(t-u)\delta X(u)du + \lambda \int_{-T}^0 K_0(t-u)\delta X_0(u)du >$$

$$> \lambda \int_0 K_0(t-u)\delta X(u)du + \varepsilon\lambda \int_{-T}^0 K_0(t-u)du$$

and using (5.41) we find out that $\delta X(t) > \varepsilon\lambda/\lambda^*$ for some small $t \in [0, t_1]$. Suppose again that we have solved the equation (5.34) for $t \in [0, t_1]$ and supplemented the prehistory interval $[-T, 0]$ to $[-T, t_1]$. Now $\delta X_0(\tau) \geq \varepsilon$ for $\tau \in [t_1 - T, t_1]$. Using the same techniques it is easy to show that $\delta X(t) > \varepsilon\lambda/\lambda^*$ for $t \in [t_1, t_2]$ where $t_2 > t_1$ and so on and distribute this property for $t \in [0, T]$.

By induction, we obtain that $\delta X(t) > \varepsilon(\lambda/\lambda^*)^n$ for $t \in [T(n-1), Tn]$, $n=1,2\ldots$, and the variations $\delta X(t)$ increase in the case of $\lambda > \lambda^*$. In other words, small perturbations about the the trivial stationary solution $X^*_S=0$ lead to increasing $\delta X(t)$ as t increases. Thus, the trivial stationary solution $X^*_S=0$ is unstable as $\lambda > \lambda^*$.

During the first part of the proof of stability, we have shown that the trivial stationary solution $X^*_S=0$ of the non-linear integral model (5.29) is stable as $0 < \lambda < \lambda^*$ and unstable as $\lambda > \lambda^*$.

Nontrivial solution.

Let us investigate the stability of a non-trivial stationary solution $X^*_S > 0$. As follows from (5.30), the second stationary solution X^*_S occurs when $\lambda \geq \lambda^*$. It can be determined from (5.30):

$$\lambda \int_0^T \overline{m}(\tau) e^{-d(\tau) - X_S{}^* \int_0^T b(\tau,\xi)d\xi} d\tau = 1 \tag{5.42}$$

or

$$\lambda \int_0^T \overline{m}(\tau) e^{-d(\tau)} e^{-X_S{}^* \int_0^T b(\tau,\xi)d\xi} d\tau = 1$$

Case 1. $\lambda > \lambda^*$; $\lambda^* - \lambda << 1$.

Comparing (5.42) and (5.31) we can notice that $X^*_S(\lambda^*)=0$ (as $\lambda=\lambda^*$) and $X^*_S << 1$ as $\lambda^* - \lambda << 1$.

To obtain the upper bound of $\delta X(t)$ we estimate $K_S(\tau) = K(\tau, t)$ first:

$$K(\tau,t) = K_S(\tau) =$$

$$= \overline{m}(\tau) e^{-d(\tau) - X^*_S \int_0^T b(\tau,\xi)d\xi} - X^*_S \int_0^T \overline{m}(u) e^{-d(u) - X^*_S \int_0^T b(u,\xi)d\xi} b(u,\tau)du$$

$K_S(\tau)>0$ as $\lambda^*-\lambda <<1$ therefore $|K_S(\tau)| = K_S(\tau)$ and

$$|K_S(\tau)| = K_S(\tau) < \overline{m}(\tau)e^{-d(\tau)-X^*{}_S\int_0^T b(\tau,\xi)d\xi} - X^*{}_S \min_{\tau,\xi} b(\tau,\xi)/\lambda$$

Then

$$|\delta X(t)| \le e^{\lambda\|K\|t}\lambda\left\|\int_0^T K(\tau,t)\delta X_0(t-\tau)d\tau\right\| <$$

$$< e^{\lambda\|\overline{m}\|t}\|\delta X_0\|\lambda\int_0^T [\overline{m}(\tau)e^{-d(\tau)-X^*{}_S\int_0^T b(\tau,\xi)d\xi} - X^*{}_S \min_{\tau,\xi} b(\tau,\xi)/\lambda\,]d\tau <$$

$$< e^{\lambda\|\overline{m}\|t}[R(\lambda) - X^*{}_S T \min_{\tau,\xi} b(\tau,\xi)]\|\delta X_0\|.$$

Using (5.31), it is easy to see that

$$|\delta X(t)| < \|\delta X_0\| \quad \text{for} \quad t\in[0,t_1],$$

$$t_1 = -\ln(1 - X^*{}_S T \min_{\tau,\xi} b(\tau,\xi))/(\lambda\|m\|) > 0$$

By analogy to case 1 for the trivial solution, we can supplement the prehistory interval $[-T, 0]$ to $[-T, t_1]$; prove the last inequality for $t\in[t_1,t_2]$, $t_2>t_1$ and distribute this property for any limit interval. It shows that a nontrivial solution X^*_S is stable in some neighborhood of $\lambda>\lambda^*$, e.g $\lambda^*-\lambda <<1$. Also it leads to the assumption that λ^* is not the only bifurcation point.

Case 2. $\lambda>>\lambda^*$.

If $\lambda>>\lambda^*$, then $X^*_S >>1$ and $K_S(\tau)<0$. Therefore $|K_S(\tau)| = -K_S(\tau)$ is determined by the second term in (5.35). As we have done before, let us construct the admissible (continuous) variation in the form (5.40) and suppose that $\Delta=o(1/X^*_S)$. Now one can see from (5.34) that

$$\delta X_0(0) = \lambda\int_{-\Delta}^{0} K_S(t-u)\delta X_0(u)du + \varepsilon\lambda\int_{-T}^{-\Delta} K_S(t-u)du \approx$$

$$\approx \varepsilon\lambda\int_{-T}^{0} K_s(t-u)du + o(X^*{}_S\Delta) =$$

$$= \varepsilon\lambda\int_{-T}^{0}[\overline{m}(t-u)e^{-d(t-u)-X^*{}_S\int_0^T b(t-u,\xi)d\xi} -$$

$$- X^*{}_S\int_0^T \overline{m}(v)e^{-d(v)-X^*{}_S\int_0^T b(v,\xi)d\xi} b(v,t-u)dv]du +$$

$$+ o(X^*{}_S\Delta) =$$

$$= \varepsilon[1 - X^*{}_S\int_0^T b(v_{mean},\xi)d\xi] + o(X^*{}_S\Delta), \tag{5.43}$$

where $v_{mean} \in (-T,0)$ is an instant obtained by applying the theorem of the mean. Since we suggest that $b(\tau, t)>0$, $\tau \in [0, T]$, $t \in [0, T]$, and $X^*_S >> 1$, then $\delta X_0(0)<0$. Moreover $\delta X_0(0) << -\varepsilon$.

Substituting the function $\delta X_0(\tau)$, $\tau \in [-T,0]$ into the linear equation (5.34) in variations implies

$$\begin{aligned}
\delta X(t) = & \\
&= \lambda \int_0^t K_S(t-u)\delta X(u)du + \lambda \int_{-T}^0 K_S(t-u)\delta X_0(u)du \approx \\
&\approx \lambda \int_0^t K_S(t-u)\delta X(u)du + \varepsilon\lambda \int_{-T}^{-\Delta} K_S(t-u)du + \\
&\quad + \lambda \int_{-\Delta}^0 K_S(t-u)\delta X_0(u)du + o(\Delta) \approx \\
&\approx \lambda \int_0^t K_S(t-u)\delta X(u)du + \varepsilon[1 - X^*{}_S \int_0^T b(v_{mean}, \xi)d\xi] + \\
&\quad + o(X^*{}_S\Delta) + o(\Delta)
\end{aligned}$$

so $\delta X(t) << -\varepsilon$ for some small $t \in [0, t_1]$ such that $t_1 = o(1/X^*_S)$.

Suppose that we have solved the equation (5.34) for $t \in [0, t_1]$ and supplemented the prehistory interval $[-T, 0]$ to $[-T, t_1]$. Now $\delta X_0(\tau) \geq \varepsilon$ for $\tau \in [t_1 - T, 0]$, $\delta X_0(\tau) \approx -\varepsilon X^*_S << -\varepsilon$ for $\tau \in [0, t_1]$. Let t increase, then

$$\begin{aligned}
\delta X(t) &= \lambda \int_{t_1}^t K_S(t-u)\delta X(u)du + \lambda \int_{t_1+t-T}^{t_1} K_S(t-u)\delta X_0(u)du \approx \\
&\approx \lambda \int_{t_1}^t K_S(t-u)\delta X(u)du + \lambda \int_0^{t_1} K_S(t-u)\delta X_0(u)du + \\
&\quad + \varepsilon\lambda \int_{t_1-T}^{t_1-\Delta} K_S(t-u)du + o(X^*{}_S\Delta) \approx \\
&\approx \lambda \int_{t_1}^t K_S(t-u)\delta X(u)du - \\
&\quad - \varepsilon X^*{}_S \int_0^T b(v_{mean}, \xi)d\xi[1 - X^*{}_S \int_0^T b(v_{mean}, \xi)d\xi]t_1/2 + \\
&\quad + \varepsilon[1 - X^*{}_S \int_0^T b(v_{mean}, \xi)d\xi] + o(X^*{}_S\Delta) \approx \\
&\approx \lambda \int_{t_1}^t K_S(t-u)\delta X(u)du + \varepsilon X^* X^* t_1 - \varepsilon X^* + o(X^*\Delta).
\end{aligned}$$

Because the function $K_S(\tau)$ changes the sign of the variation $\delta X(.)$ again $\delta X(t)$ increases for $t\in[t_1,t_2]$ and becomes $\delta X(t_2)>0$ where $t_2>t_1$ such that $t_2=O(1/X^*_S)$. When the time t increases further, we obtain $\delta X(t_3)>>\varepsilon$ where $t_3>t_2$ and so on.

So, $\delta X(t)$ changes its sign and $|\delta X(t)|$ increases for $t\in[0, T]$. One can distribute this property for any interval. It means that the nontrivial stationary solution X^*_S becomes unstable as $\lambda>>\lambda^*$, and there should be another bifurcation point λ^{**}.

Summarizing results proven in this section, we can state that the nonlinear integral model (5.29) has a bifurcation at $\lambda=\lambda^*$ where the parameter value λ^* is determined from the equality (5.31). There are three branches $X(\lambda)$ of the stationary solutions of the model (5.29) in some neighborhood of the value λ^*:

- one branch at $0<\lambda<\lambda^*$ - stable trivial solution X^0_S,
- two branches at $\lambda>\lambda^*$ - unstable zero solution $X^0_S=0$

  stable positive solution $X^*_S>0$.

Moreover, the positive stationary solution $X^*_S>0$ are stable for $\lambda^*<\lambda<\lambda^{**}$ and unstable for $\lambda>\lambda^{**}$, where λ^{**} is the second critical value of the parameter λ.

The qualitative dynamics of solutions on the plane (λ,X) is illustrated on the Figure 5.2.

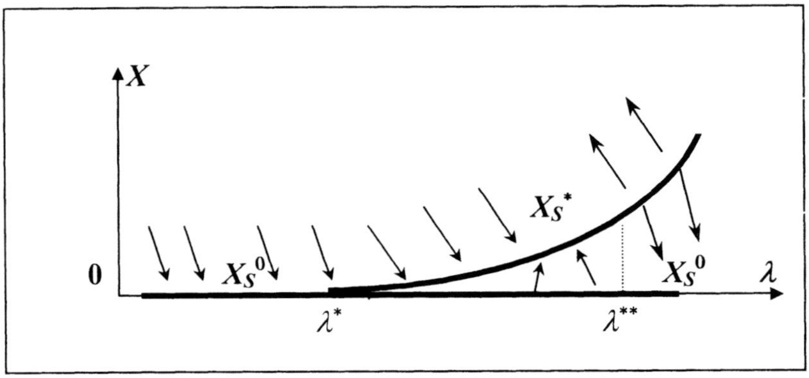

Figure 5.2. Stationary solutions X_S and bifurcation points λ^*, λ^{**} of the nonlinear integral model (5.29).

Remark. Some estimation for the second critical value λ^{**} follows from (5.43):

$$X^{**}{}_S \int_0^T b(v_{mean}, \xi) d\xi \approx 2, \qquad 0 < v_{mean} < T,$$

and the equation (5.42) for X^{**}at a given value λ^{**}:

$$\lambda^{**} \int_0^T \overline{m}(\tau) e^{-d(\tau) - X_S^{**} \int_0^T b(\tau,\xi) d\xi} d\tau = 1.$$

At least, when $b(\tau, t) \equiv b(t) > 0$, $\tau \in [0, T]$, $t \in [0, T]$, does not depend on τ, we obtain from the last two formulas and (5.31) that $R(\lambda^{**}) \approx e^2$, hence $\lambda^{**} \approx R^{-1}(e^2)$ or

$$\lambda^{**} \approx e^2 / \int_0^T \overline{m}(\tau) e^{-d(\tau)} d\tau . \tag{5.44}$$

A natural question appears whether the model (5.29) possesses more bifurcations (when λ increases). As it will be shown in the next section, the answer is positive.

2.2 Connection with Difference Models

To analyze possible further bifurcations in the integral models, we will consider two special cases of model (5.29) and its connection with difference equations. The difference equations represent a simpler mathematical tool and in many cases it is possible to establish clear mathematical properties for dynamical behavior of their solutions.

2.3.1 Single Seasonal Reproduction

By analogy with difference equations, we can expect having more bifurcations in integral models of type (5.29). To verify this statement, let us assume that in the intergal population model (5.29)

$$\overline{m}(\tau) = M\delta(\tau - 1), \quad b(\tau, \xi) = B\delta(\xi - 1), \quad T > 1, \tag{5.45}$$

where M, B =const, $\delta(t)$ is the Dirac delta-function. This choice of the given model fucntions corresponds to the case of single seasonal reproduction of individuals. Then, model (5.29) reduces to the difference equation

$$X(t+1) = \lambda M e^{-d(1)-BX(t)} X(t), \qquad (5.46)$$

i.e. the equation (5.18) with $f(X) = AXe^{-BX}$ (where the new parameter $A=\lambda Me^{-d(1)}$ is directly proportional to λ), or the Ricker model discussed in Section 1.5. It was mentioned, that when the parameter is less than the first bifurcation point then there is only one trivial solution, and it is attracting. As the parameter takes its values from the second interval, located between the first and the second bifurcation points the trivial solution becomes repelling and a new nontrivial attracting solution appears. It proves that there are more than one bifurcation points at least in the special case of (5.29), (5.45). Moreover we may anticipate more points since, as shown in Section 1.5, as the parameter increases the nontrivial solution becomes repelling and gives birth to the periodic solution with period 2. Then, cycles with increasing periods are emerged from the previous ones. It leads to a chaotic motion.

Therefore, our assumption about the existence of the second bifurcation point λ^{**} is true for the case of a single seasonal reproduction (5.29), (5.45).

2.3.2 Double Seasonal Reproduction.

Suppose that individuals in a population have two reproduction periods during their lifetime (rather than one as in previous example). Then, we can separate the population into two age classes with the numbers $X(t)$ of "younger" individuals and $Y(t)$ of "older" ones. For simplicity sake, we will suppose that the maximum age is T=2, and the reproduction ages are equal to 1 and 2 (say, each summer). Then $Y(t+1)=X(t)$ and the corresponding functions in the intergal population model (5.29) are of the form:

$$m(\tau)=M_1\delta(\tau-1)+M_2\delta(\tau-2), \quad b(\tau,\xi)=B_1\delta(\xi-1)+B_2\delta(\xi-2). \qquad (5.47)$$

If we choose the birthrate constants M_1 and M_2 and the "crowding constants" B_1 and B_2 to be the same for both ages, then the model (5.29) is reduced to the following difference equation of the second order:

$$X(t+1) = Ae^{-B[X(t)+Y(t)]}[X(t)+Y(t)], \qquad Y(t+1)=X(t) \qquad (5.48)$$

(the constant $A = \lambda Me^{-d(1)}$ as in the equation (5.46)). The analysis of this equation (Oster, 1978; Casti, 1989) shows that its dynamical behaviour is different from the one-dimentional case (5.46) and even more complicated :

(1) As for one-dimensional equations (5.18) and (5.46), there are the bifurcation values $A_0=1/2$ and A_1 such that the equation (5.48) has the only zero stable fixed point

$$Y=X=0 \text{ for } 0<A<A_0$$

and a stable positive fixed point

$$Y^*=X^*= \ln(2A)/(2B) \text{ for } A_0\leq A<A_1.$$

(2) At $A=A_1$ the stability of X^* breaks down and two cycles of period 3 appear (one is stable and another is unstable) around the fixed point (X^*,Y^*) which, however, remains locally stable for $A_1\leq A<A_2$. When A tends to A_2, the unstable inner 3-cycle decreases and strives to the fixed point that becomes repelling at $A=A_2$. The second 3-cycle remains stable at $A_2\leq A<A_3$ and splits into two cycles of period 6 at $A=A_3$. With the further growth of the parameter A, similar period-doubling bifurcations occur creating the cycles of periods $3\cdot 2^k$, k=2,3,..., and at $A=A_\infty=\lim A_k$ the dynamics becomes aperiodic and chaotic.

(3) However, at some point $A^{\{4\}}>A_\infty$ the chaos is replaced with a stable cycle of period 4. When the parameter A is further increased, this cycle generates a sequence of the cycles with periods $4\cdot 2^k$, k=2,3,..., and ends up in another chaotic regime. It is later replaced with a stable cycle of period 5, and so on.

(4) Thus, instead of one passage from stability to chaos as in a one-dimensional case, we obtain a sequence of bifurcation values

$$A_0<A_1=A^{\{3\}}<A^{\{3\}}{}_\infty<A^{\{4\}}<A^{\{4\}}{}_\infty<A^{\{5\}}<A^{\{5\}}{}_\infty<A^{\{6\}}<A^{\{6\}}{}_\infty<\ldots$$

that corresponds to the alternating ranges $\{A^{\{n\}}, A^{\{n\}}{}_{\infty}\}$, $\{A^{\{n\}}{}_{\infty}, A^{\{n+1\}}\}$ of chaotic and $n \cdot 2^k$-periodic regimes, n=3,4,5,... . In so doing, the width of the "periodic ranges" $\{A^{\{n\}}, A^{\{n\}}{}_{\infty}\}$ quickly decreases, so the chaotic dynamics prevails for high values of the parameter A. We have either periodic or aperiodic trajectories, but not both for a given value A.

The case of a double seasonal reproduction (5.47), (5.29) also provides evidence in strong support of the second bifurcation point λ^{**} in a general case of the nonlinear integral model (5.29).

2.3 Open Problems

In the previous section, we have investigated a nonlinear integral model, found stationary solutions and analyzed their stability; provided bifurcation analysis and established bifurcation values of a parameter. Also, we have found a surprising result about existing the second bifurcation point. As the comparison with difference population models shows, it may be even more bifurcations and occilations in the nonlinear integral model. Finding general conditions for such behaviour is an open problem.

More exactly, in Section 2.1 we have recognized that the first bifurcation point λ^* of nonlinear integral model (5.29) is found from (5.31)-(5.32) and describes a situation when the biological potential of a population $R(\lambda)\approx 1$. The described qualitative picture of bifurcation regimes near the first critical value λ^* is also valid for the nonlinear differential model (5.14). However, as it has been shown in Section 2.2, the non-linear integral model (5.29) has at least one more substantial change of behaviour when a positive non-trivial stationary solution becomes unstable. The value of the second bifurcation point λ^{**} has been estimated in the same section. Later, in Section 2.3, it has been shown that at least two special cases of nonlinear integral model (5.29) have more than one bifurcation points. As we have seen, integral models represent a more general tool as compared with difference models. They may produce various types of the difference models under different assumptions that lead to different pictures (patterns) of the dynamics of their trajectories. Nevertheless, these patterns have important common features exposed in the following summary:

At least in two cases (5.45) and (5.47), the nonlinear integral model (5.29) has stable oscillation regimes after the second critical value λ^{**} when a positive non-trivial stationary solution becomes unstable. If the parameter λ grows further, the model has more bifurcations when the oscillations become unstable, the period of the oscillations increases, and periodic and quasi-chaotic regimes may appear.

The analytical proof of such dynamics for the integral model (5.29) in a general case is an open problem and depends on the development of corresponding investigation techniques for the non-linear Volterra integral equations.

Another interesting open question is connected with controllability and control problems for integral models of population dynamics. Such problems belong to optimal control problems for the dynamical systems governed by integral equations. This field of optimal control is not new. General investigation techniques for such problems were studied by many authors and are discussed below in the Section 3.1 of Chapter 6.

However, optimal control of ecological systems has important specific features. Let us briefly consider them for the above age-structured models of single species population. We can introduce the control function $u(\tau,t)$ that is the specific part of the individuals of age τ harvested (taken away from the population) at the instant t into the above nonlinear differential model (5.14) or integral model (5.15) with intra-species competition.

From the pure economic viewpoint, a problem of *optimal exploitation of the population* can be formulated as the following: to find the sought-for control function $u(\tau,t)$, $\tau\in[0,\tau_{\max}]$, $t\in[0,T]$, which maximizes the consumption of individuals on some time interval $[0,T]$:

$$J(u)=\int_0^T\int_0^{\tau_{\max}}u(\tau,t)x(\tau,t)d\tau dt\to\max, \qquad (5.49)$$

under the some restrictions on the control $u(.)$: $0\le u(.)\le 1$. A preliminary analysis of this problem shows that under a sufficiently prolonged exploitation (for $T>\tau_{\max}$) the final state of population will be zero: $x(\tau,T)\equiv 0$, $\tau\in[0,\tau_{\max}]$, i.e. all the individuals will be consumed. Therefore, a meaningful problem of the population optimal exploitation should include some additional conditions of a sustainable development. It is naturally to assume that the population has to ensure the reproduction of the individuals. In view of Section 2, it means that the *biological potential* R determined by the formula (5.32) should be not less than 1.

Let us consider a problem of *stationary (stable) exploitation* as the maximization of the functional (5.49) in the age-dependent population model with the additional condition $R(u) \geq 1$ of population self-reproduction. Then, a preliminary analysis of this problem shows that the optimal population biological potential $R=1$. Thus, the strategy of a 'rational' population exploitation keeps the population at the second (non-trivial) bifurcation point analyzed in Section 2. In this case, a random perturbation of the environment or a negligible violation of exploitation conditions can lead to $R<1$. Then the nontrivial equilibrium state of the population becomes unstable, whereas the trivial state becomes stable. In other words, an irreversible mortality process might start in the population.

One can see that an optimal exploitation of real-world biological populations requires more sophisticated special models and criteria for considering feedback and ensuring the population stability against random perturbations. Such models for *in vivo* populations were thoroughly studied in (Clark, 1985).

Artificial cultivation systems for microorganism populations represent another example of controlled biological populations, which is of growing importance in modern bioengineering. These systems include communities of natural microorganisms in an artificial environment with special technical control devices. They are often used in microbiology, medicine, pharmacology, biological water purification and other industries. Corresponding population models describe main features of population development such as birth and death processes, influence of food resource, and others. Control influences include the structure and input rate of feeding substratum, the regulation of a population density by removing a part of micro-organisms, and so on. Control problems have essential peculiarities as compared with the control problems for natural plant and animal communities. Artificial cultivation systems are usually based on the *principle of continuous operation* under constant population dynamics. This principle leads to a new type of control problems for ensuring steady-state operations under random perturbations and involves analysis of steady-state trajectories stability and robustness.

Chapter 6

Modelling Of Technological Renovation In Production Systems

This chapter is devoted to modelling and optimization of industrial systems (*production systems*). Such systems belong to engineering systems and consist of various smaller *production elements* (equipment). Production systems produce products (output) and require expenses for their maintenance. Modern economic growth is characterized by structural changes based on the introduction of new technologies into the production elements. The replacement and renovation of technologies in industries undergoing technological change is one of the key aspects of economic development. However, the modelling and control of the *useful life* (*lifetime, service time*) *of technologies* has received rather little attention, in contrast to other aspects of technological progress.

This chapter concentrates on the modelling the technological renovation in production systems, which is understood as the rational replacement (renovation) of industrial equipment under conditions of technological change. This requires taking the useful life and age structure of the equipment into account. From this point of view, the modelling of equipment replacement is similar to the models of biological populations, considered in Chapter 5. The mathematical problems considered in this chapter deal with optimal control of dynamic systems governed by non-linear Volterra integral equations with sought-for integration limits.

Section 1 briefly presents the traditional approaches to modelling technological renovation in production systems. In Section 2 several integral models with the endogenous useful life of equipment are constructed and compared with other investigations. Optimization problems in these models are investigated in Section 3. Asymptotic analysis reveals interesting general patterns of the rational useful life (lifetime) of capital equipment in production systems under technological change. Under certain simplified assumptions, turnpike properties are established for the optimal useful life of equipment. Such properties are well known for other (non-integral) models of mathematical economics and indicate some basic tendencies, laws of economic dynamics. Section 3 also presents the results and discusses some of the residual problems. Detailed mathematical reasoning and proofs are contained in Section 4.

1. TRADITIONAL MODELS OF TECHNOLOGICAL RENOVATION

Models of production systems are similar to models of physical systems in many respects. They are also based on balance conservation laws, such as the balance of production and the balance of resources. These balances can be written separately for individual products or production units, and then combined in accordance with the specifics of the production process. There are many well-known production balance models; the most famous example is the Leontief input-output model (Leontief, 1986; Makarov, Levin, and Rubinov, 1995; Anthony and Biggs, 1996; etc.).

In general, modelling of the rational useful life and replacement of industrial equipment under technological change is not treatted in the traditional models of production systems. Equipment useful life has rarely been treated formally as part of a more general industrial theory. One unresolved problem is that of establishing rational strategies for capital equipment replacement under various assumptions for the behavior of technological change.

In the below treatment, *technology* is mainly connected with some amount of industrial equipment that has certain technological characteristics (throughput, price, the year of production, operating cost, etc.) that vary in time. The *useful life of technology* is understood as the service time (lifetime) of the corresponding piece of equipment, i.e. the time period between putting the equipment into operation and scrapping it.

In contrast to classical renewal theory, which is based on stochastic models, the models proposed below are deterministic, emphasizing optimization of the unknown *useful life* of technologies. The models introduce a new endogenous (sought-for) time threshold for the use of technology (equipment): equipment created before this time is obsolete and is not used at the current time.

Before constructing integral models for equipment useful life optimization, the authors briefly consider some traditional approaches to modelling of equipment replacement. They include:

- aggregated phenomenological models and optimal investments control;
- models of equipment age structure;
- statistical modelling of equipment renovation and replacement.

1.1 Aggregated Models of Optimal Investments

Aggregated models are based on hypotheses about the behaviour of a production system. The authors consider only the simplest of these models - a production system that is assumed to be autonomous (independent in time) and produces a uniform aggregated product.

The simplest model assumes that, without any investment, the production rate (throughput) of a production system (production facility) decreases at a rate proportional to the production rate (exponential decay). However, it is possible to invest into the production facility in order to increase the production rate. The model is described by the following single linear ODE:

$$dY/dt = -\alpha Y + \gamma I, \qquad \alpha,\ \gamma = \text{const}>0 \tag{6.1}$$

with respect to the production rate $Y(t)$, $t\in[t_0, T]$. Here the investment rate $I(t)$, $t\in[0, T]$, is positive and bounded:

$$0 \le I(t) \le I_{\max}, \qquad t\in[0, T],$$

and may be considered as a controlling function.

Let us also assume that the total cost of the capital equipment is proportional to its production rate. In this case, the problem of determining the optimal investment plan for the production facility for time horizon $[t_0,T]$ is defined as finding the optimal *investment plan* $I^*(t)$, $t\in[t_0, T]$, which maximizes the following objective function:

$$J(I) = \beta Y(T) + \int_0^T [Y(t) - I(t)]dt \rightarrow \max \tag{6.2}$$

under the equality constraint (6.1).

The $\beta Y(T)$ component of (6.2) determines the *salvage value* of the equipment at time T (if production is stopped at time T). If the facility were to operate beyond T without any further investment, it could produce

terminal revenue of $\int_0^{\infty} Y(t)dt = (1/\alpha)Y(T)$.

Such optimization models represent a common tool for macroeconomic modelling. The optimal control problem (6.1)-(6.2) may be investigated using the *Pontryagin's maximum principle* (Pontryagin et al, 1962). Its behaviour is typical for such linear dynamic systems and includes the following features (Luenberger, 1979):

a) The optimal control $I^*(t)$ appears to be bounding: $I^*(t)=0$ or $I^*(t)=I_{max}$ for all $t\in[t_0, T]$, and may have a switch (inversion) point at

$$t_s = T - \ln[(1/\alpha-\beta)/(1/\alpha-1/\gamma)].$$

b) If $\gamma < \min(\alpha, 1/\beta)$ (i.e., investment efficiency γ is low), then the optimal policy means no investment at all ($I^*\equiv 0$).

c) If $1/\beta \leq \alpha < \gamma$ (i.e., the salvage value is greater than the terminal revenue), then the optimal policy is to invest at the maximum rate I_{max} for the entire period $[t_0, T]$.

Thus, the qualitative picture of the optimal trajectories is too restrictive and does not have interior regimes in the domain $0 \leq I(t) \leq I_{\max}$ (similar to the golden rule of economic growth).

This approach is similar to the aggregate models of macroeconomics. The first obvious improvement to the model (6.1)-(6.2) would be to use two-factor production functions and include exogenous factors like technological change, managerial decisions, etc. (Hritonenko & Yatsenko, 1996). However, it is even more critical to consider the internal structure of the the production facilities. In particular, the different useful lives of production elements are important for the modelling of equipment replacement in industrial systems. This leads to another type of mathematical models, considered below.

1.2 Age-Specific Models of Equipment Replacement

The age-distribution of production equipment is one of the key endogenous factors influencing production dynamics. Since expenditures for the maintenance and repair of *equipment units* depend on the equipment age, rational investment policy requires taking the age structure of the production equipment into account.

Models of equipment age structure are usually described with partial differential equations (PDEs). A simple model of this is provided by the following PDE (*evolutonary equation*):

$$\partial x(t,\tau)/\partial \tau + \partial x(t,\tau)/\partial t = -\alpha(t,\tau)\,x(t,\tau), \tag{6.3}$$

$$x(t,0) = m(t),$$

$$x(0,t) = \varphi(\tau),$$

where

$x(t,\tau)$ is the *equipment distribution density* at time t in accordance with the equipment age τ,

$\alpha(t,\tau)$ is the *failure coefficient* for the equipment unit of age τ at time t,

$m(t)$ is the quantity of the *equipment units* (EUs) introduced into operation at time t,

$\varphi(\tau)$ is a given initial equipment age distribution at time t=0.

The evolutionary equation (6.3) is often used in modelling the age structure of biological populations. Derivation of equation (6.3) is considered above in Chapter 5 (model (5.9)). An integral model equivalent to the differential model (6.3) can be treated like the integral model described in Chapter 5 for ecological systems.

The total quantity of operating EUs is

$$P(t,Z) = \int_0^Z [1 - f(t,\tau)]x(t,\tau)d\tau, \tag{6.4}$$

where $f(t,\tau)$ is the percentage of the EUs of age τ being repaired at time t, and the constant Z>0 is the EU service time (it may be considered a control parameter).

The PDE model (6.3)-(6.4) can be used for determining the optimal investment plan for the production facility. The total expenditures on operating a production system on planning horizon $[0,T]$ can be described as

$$S(t,Z) = \int_0^T \Big(p(t)m(t) + \int_0^Z x(t,\tau)[\gamma(t,\tau) +$$

$$+ r(t,\tau)f(t,\tau) + q(t,\tau)(1 - f(t,\tau))]d\tau\Big)dt, \tag{6.5}$$

where $p(t)$ is the cost of one new EU at time t, and $\gamma(t, \tau)$, $r(t, \tau)$, $q(t, \tau)$ are the *specific expenditures*, respectively, for spare parts, repairs, and maintenance per one EU of age τ at time t. The unknown (sought-for) controls could include the quantity $m(t)$ of new EUs, the EU service time Z, and other endogenous parameters.

One can notice that such differential models describe the constant equipment useful life (maximum service time). If the equipment useful life is variable in time, then it is convenient to use the integral dynamical model with variable memory considered in Sections 2 and 3.

1.3 Statistical Models of Equipment Renewal

The most thoroughly developed tools for modelling of equipment replacement are based on the statistical methods (Karlin, 1968; Tijms, 1986; and others). These tools make it possible to calculate the required amounts of equipment replacement by using statistical data about possible equipment failures. In this approach, the equipment parameters are usually assumed to be independent of the current time, and technological progress is excluded. The authors briefly illustrate this approach with an example of the famous *renewal equation* statistical model.

Let us consider a production system with many equipment components (elements). It is assumed that reliable experimental results show that the probability of failure of a critical component is $f(\tau)d\tau$ during time interval $d\tau$. The number of these critical components is equal to N, and we want to determine the expected number of component renewals that will be needed by time t.

We define *renewals of the first kind* as renewals of the initial components, *renewals of the second kind* as the renewals of renewals of the first kind, and so on. Then, if $u_i(t)$ is the expected number of renewals of the i-th kind at time t, we have

$$u_1(t) = Nf(t), \qquad u_i(t) = \int_0^t u_i(t-\tau)f(\tau)d\tau, \qquad i = 1,2,...$$

The expected total number $U(t)$ of renewals by time t is given by

$$U(t) = \int_0^t \sum_{i=1}^{\infty} u_i(\xi) d\xi =$$

$$= N \int_0^t f(\xi) d\xi + \int_0^t d\xi \int_0^{\xi} \sum_{i=2}^{\infty} \int_0^t u_i(\xi - \tau) f(\tau) d\tau =$$

$$= NF(t) + \int_0^t U(t - \tau) dF(\tau), \quad \text{where} \quad F(t) = \int_0^t f(\tau) d\tau$$

The last equality represents the Volterra integral equation of the second kind for the total number of renewals $U(t)$

$$U(t) = NF(t) + \int_0^t U(t - \tau) dF(\tau) \tag{6.6}$$

which is known as the *renewal equation.* Many researchers have studied its remarkable properties. The interested reader may confer the numerous statistical works (Karlin, 1968; Tijms, 1986; and others) for more detail. The basic renewal scheme considered above has been generalized for the case of multi-unit systems, minimum of repair costs, waiting times, restricted availability of repair and maintenance, and other aspects (Sheu, 1991; Agrafiotis & Tsoukalas, 1990; and so on).

2. MODELS OF EQUIPMENT REPLACEMENT UNDER TECHNOLOGICAL CHANGE

In this section, the authors combine the endogenous growth models of a separate firm (plant, enterprise) with the optimization techniques developed in integral macroeconomic models to obtain insights into the rational scrapping of obsolete industrial equipment in production systems. The problem of optimal equipment replacement is analyzed in the framework of a self-organizing firm's behavior in market economics with intensive technological change.

Mathematical modelling of evolutionary economics under technological change (TC) has been rigorously examined by many authors during the last decades. A perspective self-organizing evolutionary approach is developed in (Silverberg et al., 1988, Englmann, 1992, Bruckner et al., 1993), which unites the aspects of new technologies diffusion, diversity, age-dependent effects, and others. The corresponding mathematical models include a certain model block that describes renovation of capital equipment (or technologies). The authors will identify this equipment renovation model and investigate it using formal methods developed in macroeconomic optimization theory (Kantorovich and Zhiyanov, 1973, Glushkov, 1977; Glushkov, Ivanov, and Yatsenko, 1980, 1982; Yatsenko, 1991; Hritonenko and Yatsenko, 1994, 1996, 1999). These methods make it possible to study the problem of optimal technological renovation on the basis of integral dynamical models.

Section 2.1 briefly outlines the basic modelling assumptions and introduces a self-organizing model of an individual enterprise in evolving market economics. An aggregate integral model with the endogenous time of equipment scrapping is derived in Section 2.2 and compared to similar investigations. The disaggregated integral model of production systems is constructed in Section 2.3. Qualitative analysis of optimal trajectories in these models is provided in Section 3. The most interesting results are related to the asymptotic behavior of optimal trajectories of technological replacement and renovation.

2.1 Self-Organizing Market Model of Enterprise Under Technological Change

The first model for the rational replacement of capital equipment in a stand-alone firm was probably proposed in (Malcomson, 1975). The problem consists of discounted maximization of the firm's profit (net revenue) that includes the output sold, the operating cost of all machines in use, and the cost of purchasing new machines under the given output, over an infinite time horizon.

It is supposed that the technological change (TC) is materialized (embodied) in newer equipment (in new vintages of the machines being operated). This description of the technological change is known as *embodied TC* and the corresponding economic models are usually referred to as *models with embodied TC*.

A dynamic self-organizing model of a firm in conditions of market competition and embodied technological change was proposed in (Silverberg, 1987) and elaborated in (Silverberg et al., 1988). This model is used below for the analysis of rational equipment renovation in production systems.

The model considers *a technology* to be some amount of *industrial equipment* with given characteristics, perhaps incompletely known in the face of uncertainty with respect to the future TC course. As outlined in

(Silverberg et al., 1988), this evolutionary model may be categorized as a set of differential-difference equations with an age-dependent effect whose mathematical properties are poorly understood even in the simplest cases.

Under certain simplified assumptions this model leads to non-trivial results, revealing in particular, turnpike properties for the optimal scrapping time of obsolete capital equipment. This analysis is carried out in the Sections 3 and 4.

Namely, the *following hypotheses are made*:

- the uncertainty of market processes is eliminated and the maximizing behavior of firms is assumed;
- the total industry-level market demand is assumed to be known;
- the influence of each separate firm on the market state is ignored;
- the future course of technological progress is assumed to be completely known.

These assumptions simplify the notion of technology that is now embodied in some given features of production inputs. Note that the model does not assume the existence of any steady-state distributions of capital equipment and is not an equilibrium model.

In this model, the technological change is embodied in vintages (versions) of the industrial equipment used, and the capital stock of the ***i***-th firm in an industry is represented as an aggregation over non-decaying vintages between the current time t and scrapping date $a_i(t)$:

$$K_i(t) = \int_{a_i(t)}^{t} m_i(\tau)d\tau, \tag{6.7}$$

where

$K_i(t)$ is the capital stock (measured in some units of productive capacity),

$m_i(t)$ is the gross investment at time t,

$a_i(t)$ is the date of scrapping the oldest technologically obsolete equipment, $a_i(t) < t$, $i=1,\dots,N$,

N is the total number of all firms in the industry.

The desired and actual scrapping dates are assumed to be equal, i.e. firms are able to finance their desired investment programs. Then the amount of the scrapped capacity is $S_i(t) = m_i(a_i(t))a_i'(t)$ and the net change of capital stock is

$$dK_i(t)/dt = m_i(t) - S_i(t). \tag{6.8}$$

Labour is assumed to be the only current cost of production and can be decomposed into prime and overhead components. The prime unit labour component is described as

$$L_i(t) = \int_{a_i(t)}^{t} l_i(\tau) m_i(\tau) d\tau / K_i(t), \tag{6.9}$$

where $l_i(\tau)$ is the *historical technological labour/output coefficient.* This coefficient is determined by the age structure of the capital stock and by the history of TC. It varies over time due to additions of more productive new equipment through investment and removal of the marginal equipment through scrapping. An overhead labour per unit output will be assumed proportional to the prime unit labour at full operating capacity and be included into the equality (6.9).

The current production output is

$$y_i(t) = u_i(t) K_i(t), \tag{6.10}$$

where the level $u_i(t)$ of production is chosen in the model to compensate for the deviation of the current delivery delay $D_i(t)$ from an industry-wide level $D_0(t)$:

$$\begin{aligned} du_i(t)/dt &= C_u(D_i(t) - D_0(t))\, u_i(t)(1-u_i^2(t)), \\ &\qquad C_u = \text{const} > 0, \quad \text{for} \quad u_i(t) < 1, \\ du_i(t)/dt &= 0 \qquad \text{for} \quad u_i(t) = 1. \end{aligned} \tag{6.11}$$

The delivery delay $D_i(t) = B_i(t) / y_i(t)$, where the order backlog $B_i(t)$ is governed by the equation:

$$dB_i(t)/dt = d_i(t) - y_i(t), \tag{6.12}$$

where the number of incoming orders $d_i(t)$ is equal to

$$d_i(t) = f_i(t) \Phi_T(t), \tag{6.13}$$

$\Phi_T(t)$ is the given *total market demand*, $f_i(t)$ is the market share of the i-th firm in the percentage of real orders, $0 < f_i < 1$, $\Sigma f_i = 1$.

The evolution of market shares is governed by the following equation:

$$df_i(t)/dt = C_s(\ E_i(t) - \langle E\rangle(t)\ )\, f_i(t), \quad C_s = \text{const} > 0, \tag{6.14}$$

where $E_i(t)$ is the *competitiveness* of the ***i***-th firm, and
$\langle E\rangle = \Sigma f_i\, E_i$ is the *average competitiveness* of all firms in the industry.

The competitiveness $E_i(t)$ is defined as a linear combination of terms reflecting price and delivery delay differentials:

$$E_i(t) = -\ln p_i(t) - C_E D_i(t)\ , \quad C_E = \text{const} > 0, \tag{6.15}$$

Finally, the firms' prices $p_i(t)$ are determined as a compromise between strict cost-plus pricing and relative competitiveness:

$$d(\ln p_i(t))/dt = C_m(\ln p_i(t) - \ln p^0{}_i(t)) +$$

$$+C_c(E_i(t) - \langle E\rangle(t)) f_i(t), \quad C_c,\, C_m = \text{const} > 0, \tag{6.16}$$

(C_c=0 in the case of pure monopoly at and C_m=0 in the case of pure competition).

Some empirical economic approaches for finding the time $a_i(t)$ of obsolete technology (equipment) replacement (the *scrapping date*) were briefly covered in (Silverberg et al., 1988). They are based on using additional parameters such as unit operating costs and target payback period. In this chapter, the authors will show that the rational scrapping date may be explicitly determined based on the rate of technological change.

The resulting endogenous production-market model (6.7)-(6.16) represents a system of non-linear integral-differential equations with after-effect. It may be solved numerically for some time interval (t_0, t_1) under the given scrapping time $a_i(t)$, the model constants, and necessary initial conditions for $t \le t_0$. However, in the present form, this model is too complicated for theoretic optimization analysis. For this reason, one more *additional assumption* is made:

Let us assume that the current delivery delay $D_i(t)$ coincides with its given industry-wide level $D_0(t)$: $D_i(t) \equiv D_0(t)$. This allows us to separate this system of equations into two simpler problems. Indeed, by (6.11), we obtain $u_i(t) \equiv$ const. For the sake of simplicity, we consider the initial condition $u_i(t_0) = 1$, then $u_i(t) \equiv 1$. Then in view of (6.12)-(6.16) one can determine $f_i(t)$, $E_i(t)$, $p_i(t)$, $d_i(t)$, $B_i(t)$, $y_i(t)$, $t \in (t_0, t_1)$, separately from equations (6.7)-(6.9), and, hence, the current production output $y_i(.) \equiv K_i(.)$ is given in the left side of equation (6.10). Now the nonlinear integral model (6.7)-(6.10) may be considered independently.

Thus, the authors have separated the price-delivery-market-competition part (6.11)-(6.16) of the self-organizing firm model (6.7)-(6.16) from its technological–production part (6.7)-(6.10). It will allow us to concentrate on the modelling of rational technological structure of the firm's productive capacities as a stand-alone object and later connect it with the surrounding market reality.

For simplicity, the firm's subscript i is omitted in the following sections.

2.2 Aggregated Model with Endogenous Useful life of Equipment

In contrast to other similar economic-mathematical works, the equipment scrapping time $a(t)$ is treated as a sought-for (endogenous) variable. To determine this endogenous variable, the authors will formulate and use an optimization problem for equipment (technology) scrapping policy in the framework of the aggregate integral model (6.7)-(6.10).

The integral model (6.7)-(6.10) may be rewritten in the following form :

$$K(t) = \int_{a(t)}^{t} m(\tau)d\tau, \tag{6.17}$$

$$L(t)K(t) = \int_{a(t)}^{t} l(\tau)m(\tau)d\tau, \tag{6.18}$$

where

$K(t)$ is the capital stock (measured in units of productive capacity),
$m(t)$ is gross investment at time t,
$a(t)$ is a scrapping time for technologically obsolete equipment,

$l(t)$ is historical technological labour/output coefficient,
$L(t)K(t) = M(t)$ is the prime labour unit coefficient.

It should be noted that the first mathematical model for optimal replacement of capital equipment in a separate firm was proposed in (J.M.Malcomson, 1975). It is described by the same equations (6.17)-(6.18).

The model (6.17)-(6.18) does not distinguish among the different types of products and technologies (equipment) of the firm and describes the products of the firm in an aggregate manner. However, it does consider the age structure of the firm's productive capacities and contains control tools for this structure. From a mathematical point of view, it presents a system of non-linear integral equations with controlled after-effect duration (controlled memory) with respect to the sought-for functions a and m (Hritonenko and Yatsenko, 1996).

The following section compares the model (6.17)-(6.18) with similar macroeconomic models.

2.2.1 Integral Macroeconomic Models of Technological Renovation

While integral models with embodied TC were initially applied to large-scale macroeconomic systems, the economic processes are similar for all levels of economic management. The models considered in this chapter are intended for a separate enterprise/firm level of economic development.

Integral economic models have made it possible to investigate the useful life (service time) of technologies and rational rates of technological renovation. These problems have been first considered with regard to other economic aspects such as technological change, presence of resources, production efficiency criteria, and so on. Importance of these problems in applications motivated investigation of the integral economic models and their use for optimization of different scale economic development.

All the integral macroeconomic models described below fall into the class of *macroeconomic models with embodied (materialized) technological change* (see also Section 3.1). The first model of this type (the *Solow integral model*) was developed by R.Solow (Solow, 1960).

The first model with controlled useful life of technologies was proposed in (Kantorovich and Zhiyanov, 1973). It is a modification of the Solow integral model, described by the following equations:

$$Q(t) = \int_{a(t)}^{t} F[\tau, k(\tau), \lambda(\tau)]d\tau, \qquad (6.19)$$

$$\Lambda(t) = \int_{a(t)}^{t} \lambda(\tau)d\tau, \tag{6.20}$$

where

$Q(t)$ is the total production output of an economic system,
$\Lambda(t)$ is the total amount of labour force used,
$k(\tau)$ is the new capital assets (productive capacities) put into operation per unit of time at time τ,
$\lambda(t)$ is the amount of labour required for the productive capacity $k(\tau)$,
$F(\tau, k, \lambda)$ is the production function for capital $k(\tau)$ and labour $\lambda(\tau)$.

The principal innovation of this model (and other models mentioned below) is the introduction of the new endogenous function $a(t)$ that is the *time limit of capital use*: the productive capacities created before the $a(t)$ instant are not used at the current time t. The value t - $a(t)$ determines by the *useful life (service time)* of the capital (capital equipment).

A similar economic model (the *integral production function*) was introduced in (Petrov & Pospelov, 1979) for modelling of developing economics. At first glance, it is different from the above models and is described as

$$Q(t) = \int_{\varphi(t)}^{\phi(t)} m(t, \chi)d\chi, \tag{6.21}$$

$$\Lambda(t) = \int_{\varphi(t)}^{\phi(t)} \chi m(t, \chi)d\chi, \tag{6.22}$$

This model makes the labour force Λ dependent on the production output Q through the given distribution $m(\tau,\chi)$ of productive capacities and the sought-for parameter $\phi(t)$ in the upper integration limit.

In model (6.21)-(6.22), the distribution $m(t,\chi)$ of operating capital is built in accordance with *the specific labor expenditure* χ of the capital unit rather than with the time τ of putting the capital into operation as in the above integral models. The model accounts for technological change by decreasing the function $m(t,\chi)$ on χ. The integrals in models (6.21),(6.22) are evaluated for independent variable χ and the limit $\phi(t)$ of integration is a sought-for variable.

It was shown in (Yatsenko, 1991) that, under certain natural economic assumptions, the substitution $\chi \rightarrow \tau$ of the independent variables χ and τ in accordance with their economic content reduces the model (6.21)-(6.22) to that of the form (6.18)-(6.19).

In fact, the integral production function (6.21)-(6.22) represents a one-dimensional version of earlier and more general integral economic models based on similar assumptions (Johansen, 1972). In the Johansen models, production units are distinguished by their specific expenditures of several independent economic factors (rather than by just the labor expenditure χ), and the models involve multiple integrals and multidimensional integral equations.

The macroeconomic integral model (6.19)-(6.20) turns out to be the foundation for many problems discussed later on. In 1977 V.Glushkov, a well-known Ukrainian scientist, proposed a two-sector integral model of macroeconomic renovation. This model represents the relationship between two *production sectors*: A - the production of production assets (productive capacities) and B - the production of consumer goods (Glushkov, 1977; Glushkov, Ivanov, & Yatsenko, 1980, 1982). Another macroeconomic model with disaggregated assets was considered in (Makarov & Rubinov, 1973); see also (Makarov, Levin and Rubinov, 1995). At that time an intensive research of integral macroeconomic models was started and significant results have been obtained, see (Hritonenko and Yatsenko, 1996, 1999) and the references therein. During this research, turnpike properties (Yatsenko and Hritonenko, 1994) similar to those discovered for other models of mathematical economics (Lancaster, 1968, McKenzie, 1973, Makarov, Levin and Rubinov, 1995) were proven for the optimal useful life of equipment. Such models were applied to the conversion of the defense industry, technological renovation in hierarchical regional economic-ecological systems, modernization of agricultural manufacturing systems, etc.

Multi-sector integral models (Hritonenko & Yatsenko, 1996) simultaneously describe the processes of equipment renovation and the distribution of operated EUs among industrial branches (production sectors or units) as in the case of linear input-output production models.

2.2.2 Statement of Optimization Problem

From the economic point of view, a natural objective is to minimize the expenditures (expenses) for firm's operations. In the model (6.7)-(6.10), these expenditures include the total current cost of production $L(t)K(t)$ and the gross investment (in production units) $P(t)m(t)$, where the given function $P(t)$ is *the unit cost of new capital equipment* (per the production unit capacity). Thus, the total expenditures of the firm on the given time interval $[t_0, T)$ are equal to

$$I = \int_{t_0}^{T} \rho(t)[L(t)K(t) + P(t)m(t)]dt, \tag{6.23}$$

where $\rho(t)$ is a discounting factor, $0<\rho(t)\leq 1$, $\rho'(t) \leq 0$.

In accordance with the previous section, the historical technological labour/output coefficient $l(t)$, the cost $P(t)$ of new equipment units, the firm's production output $y(t)=K(t)$, and the discounting multiplier $\rho(t)$ in the intergal model (6.7)-(6.10) may be considered as the given characteristics for $t\in[t_0, T)$.

We choose the gross investment (in the capacity units) $m(t)$ and the time $a(t)$ of obsolete capital equipment (technology) scrapping as sought-for control variables, $t\in[t_0, T)$. Then the function $L(t) = t-a(t)$ describes the sought-for useful life of the equipment.

Now we are going to impose some restrictions for the sought-for variables. First of all,

$$0 \leq m(t) \leq M(t), \tag{6.24}$$

where the maximum possible investment $M(t)$ is determined by of the firm's financial constraints. It is also natural to assume that $a'(t) \geq 0$, i.e. scrapped capacity cannot be used again. Finally, as the model (6.7)-(6.10) now stands, a specific age structure of the firm's equipment is known and defined by the investments $m_0(t)$ made during some former time period (*pre-history of the process*) $[a(t_0), t_0]$:

$$m(\tau) = m_0(\tau), \quad \tau \in [a(t_0), t_0]. \tag{6.25}$$

Thus, the mathematical statement of the above-mentioned optimization problem (OP) consists of determining the functions $m(t)$, $a(t)$, $t\in[t_0, T)$, $T\leq\infty$, which maximize the objective function (6.23) under the equality constraint (6.17), the inequalities constraints (6.24) and $a'(t) \geq 0$, $a(t) < t$, and the initial conditions (6.25). The qualitative analysis of this optimization problem is provided in Section 3.

2.3 Disaggregated Integral Model of Equipment Replacement

In this section, a disaggregated multi-product integral model is constructed to describe equipment replacement and modernization for production systems under real-life technical and operational conditions. It takes into account such important features of real manufacturing systems as implementation of technological operations at specific calendar periods, standard schedules of the operations, as well as basic and auxiliary equipment. Such models open up new possibilities in simulation and optimization of the development of complex manufacturing and engineering systems.

2.3.1 Description of Production System

Let us start with a brief description of the basic modelling assumptions.

In a production (manufacturing) system, we will distinguish the system's *basic elements* from its *auxiliary elements*. A *basic element* is understood to be an element (an equipment unit) of the system capable to perform certain functions either independently or in conjunction with some auxiliary elements. An *auxiliary element* is understood to be an element capable of functioning only together with a basic element. A subsystem that consists of a single basic element and a set of associated auxiliary elements is referred to as *an assembly*.

We also consider manufacturing processes that need to be implemented during *specific calendar periods* (for example, numerous technological processes in agriculture, construction industry, etc.) in accordance with a specified sequence of technological operations during the annual production cycle. An annual production cycle of operations should be subdivided into calendar periods, which we refer to as *design periods*. A *design period* is understood to be a time interval during which none of the continuing operations terminates and during which no new operation starts.

Next, specified standard schedules of the technological operations are performed among the design periods. If a certain operation takes several periods, the schedule of this operation is divided in proportion to the length of the periods. The demand for equipment is determined on the basis of peak periods that require the maximum quantity of equipment units. For this purpose, *an equipment allocation problem* should be solved with respect for the equipment units available at a certain time, for each type of operation, and for each design period. Well-known scheduling algorithms and programs may be used to solve this problem. As result, an annual schedule of equipment utilization is obtained, and the maximum demand for each type elements is determined during some definite periods (*peak periods*). After solving the allocation problem, we may investigate rational equipment replacement for the peak periods on the basis of the integral dynamic model proposed below.

2.3.2 Construction of Model

Let us set up some initial definitions:

- I is a specified set of technological operations during the design year; i is the index of the operation, $i \in I$, $I=|n|$;
- J is the set of the types of basic elements; j is the index of basic element types, $j \in J$, $J=|N|$;
- L is the set of types of auxiliary elements; α is the index of types of auxiliary elements, $\alpha \in L$;
- A_{ij} is a vector that determines the composition of the j-th assembly for implementation of the i-th operation, $A_{ij}=(\lambda_i, \lambda_{\alpha 1},\ldots, \lambda_{\alpha i})$;
- λ_i is the number of basic operations of type j, usually $\lambda_i = 1$;
- $\lambda_{\alpha i}$ is the number of auxiliary operations of type α_r, $r=1,\ldots,S$, in the assembly A_{ij} carried out by the basic elements of type j.

We assume that *assembly* A_{ij} with the pair (i, j) is uniquely determined. Generally speaking, different auxiliary elements may be assembled to implement the i-th operation with the type j basic elements; that is, $A_{ij} \neq A'_{ij}$, where $A'_{ij}=(\lambda_i, \lambda'_{\alpha 1},\ldots, \lambda'_{\alpha i})$. However, we will assume that on the basis of a specific criterion (i.e., productivity), a single assembly A_{ij} has been chosen from a possible list of such assemblies. Next, we can consider in place of the assembly A_{ij} a type j basic element that implements the i-th operation with uniquely determined productivity β_{ij}. Now the problem of distributing the production elements among operations may be solved in two steps: (a) determining the required quantity of the basic elements, and (b) calculating the required quantity of auxiliary elements of type α, $\alpha \in L$ (by multiplying by the corresponding λ_α).

To construct a model that describes real-life production processes we introduce the following definitions:

- k is the period index, $k \in K$;
- τ is the year when the basic element has been incorporated into the manufacturing system;
- $[t_0, T]$ is the planning design interval, measured in years;
- $m_{ij}(\tau)$ is the quantity of basic elements of type j that were incorporated into the manufacturing system in year τ and perform the i-th operation;
- $\beta_{ij}(\tau,t)$ is the productivity of an assembly containing the type j basic elements that were incorporated into the production system in year τ and are performing the i-th operation in year t;
- $m_{ij}(\tau)$ is manpower (the number of personnel) responsible for the operation of a single assembly with the basic element of type j incorporated in year τ, which implement the i-th operation;
- $a_j(t)$ is the scrapping time of the obsolete basic equipment of the type j, i.e., if a specified basic element went into operation in year τ, it is in service while $\tau > a_j(t)$ and is written off once $\tau = a_j(t)$;
- $b_i(t)$ is the specified standard schedule of i-th operation during the design period of the production cycle in year t,
- $B(t)$ is the total manpower necessary for operation of the whole production system in year t.

The equipment demand is determined from the condition of guaranteed implementation of standard schedule for all specified operations during each design period of year t, $t \in [t_0, T]$:

$$b_i(t) = \sum_{j=1}^{N_i} \int_{a_j(t)}^{t} \beta_{ij}(\tau,t) m_{ij}(\tau) d\tau, \quad i = 1,\ldots,n_k, \tag{6.26}$$

where

N_i is the number of basic elements of different types used in the i-th operation,

n_k is the number of different operations used during design period k,

k is the fixed period index, $k \in K$.

The demand for manpower that guarantees the performance of the standard schedule is determined from the following equation:

$$B(t) = \sum_{i=1}^{n_i} \sum_{j=1}^{N_i} \int_{a_j(t)}^{t} q_{ij}(\tau) m_{ij}(\tau) d\tau. \tag{6.27}$$

The model (6.26)-(6.27) represents a system of Volterra integral equations of the first kind with controllable memory. It describes the process of replacement and modernization of the equipment elements of a production system, by taking the technological change into account. The technological change causes the increase of the productivity functions $\beta_{ij}(\tau, t)$ in the variable τ (the year when the equipment unit is placed into operation).

The functions $b_i(t)$, $B(t)$, $\beta_{ij}(\tau, t)$, and $q_{ij}(\tau)$ are always specified in advance, $\tau\in[\tau_0,T]$, $t\in[t_0,T]$. The functions $m_{ij}(t)$, $i=1,\ldots,n_k$, $j=1,\ldots,N_i$, are usually unknown, and the functions $a_j(t)$, $j=1,\ldots,N_i$, may be either specified in advance or they may be unknown, depending on the formulation of the problem.

Equations (6.26)-(6.27) depend on the pre-history $[\tau_0, t_0]$ of the production process. Therefore, the functions $m_{ij}(\tau) \equiv m^0_{ij}(\tau)$, $\tau\in[\tau_0, t_0]$, along with the values $a_j(t_0) \equiv a^0_j<t_0$, $i=1,\ldots,n_k$, $j=1,\ldots,N_i$, have to be specified. The functions $m^0_{ij}(\tau)$ are determined by solving the allocation problem mentioned in subsection 2.3.1 for time t_0. The values of $a_j(t_0) \equiv a^0_j$ are known from the real process of writing off obsolescent equipment units of the production system over the past history $[\tau_0, t_0]$.

Depending on the number of unknown functions and the number of equations in the model (6.26)-(6.27), we may consider one of two basic problems. If the number of equations is equal to the number of unknowns and the problem is completely determined, we have a prediction problem of the development of a manufacturing system. If there are more unknowns than equations, it leads to an optimal control problem in the model (6.26)-(6.27).

2.3.3 About Prediction Problems

The model (6.26)-(6.27) usually contains $\sum_{i=1}^{n_i} N_i$ sought-for functions $m_{ij}(t)$, more than one unknown $a_j(t)$, and n_k+1 equations. In the case $N_i>1$, the number of unknowns in the model is larger than the number of equations. Here, we assume that, after solving the allocation problem and determining the peak periods $k=k_i^{\max}$ for $j\in J$, each obtained effective set of utilized equipment for i-th operation, $i=1,\ldots,n_k$, consists of a single type j of basic elements, that is, $N_i = 1$, $i = 1,\ldots,n_k$.

When $N_i = 1$, the integral model (6.26)-(6.27) assumes the following form:

$$b_i(t) = \int_{a(t)}^{t} \beta_i(\tau,t) m_i(\tau) d\tau, \quad i = 1,\ldots,n_k, \qquad (6.28)$$

The prediction problem with control of the replacement of equipment

$$B(t) = \int_{a(t)}^{t} \sum_{i=1}^{n_i} q_i(\tau) m_i(\tau) d\tau. \qquad (6.29)$$

elements, i.e., with the sought-for function $a(.)$, represents an important applied problem. In this case, the system (6.28)-(6.29) contains n_k+1 equations and n_k+1 unknowns $m_i(.)$, and $a(.)$, and is a system of Volterra integral equations of the first kind with the sought-for lower limit of integration.

The solvability of equations (6.28)-(6.29) has been investigated in (Golovach & Yatsenko, 1992; Hritonenko & Yatsenko, 1996), where it is shown that the system (6.28)-(6.29) has the a unique continuous solution $m_i(t)>0$, $i=1,\ldots,n_k$, $a(t)<t$, $t\in[t_0,T]$, under pretty natural conditions. These conditions include positiveness and differentiability of the functions $b_i(t)$, $P(t)$, $\beta_i(\tau, t)$, and $q_i(\tau)$, the presence and continuity of the technological progress (i.e., $\partial\beta_i(\tau, t)/\partial\tau > 0$ for all i), the presence of the deterioration of equipment (i.e., $\partial\beta_i(\tau, t)/\partial t < 0$, $dq_i(\tau)/dt < 0$ for all i), and similar. Proof of this result employed a technique based on the *contractive mapping principle* and developed for the integral dynamical models with controllable memory in (Yatsenko, 1991).

Effective numeric algorithms for solving equations (6.28)-(6.29) and similar were constructed in (Yatsenko, 1991; Aistrakhanov & Yatsenko, 1992; and others). Next necessary step in solving the prediction problem requires the aggregation of its solution. The result of solving the equations (6.28)-(6.29) for each peak period $k=k_i^{\max}$, $j\in J$, is the predicted quantity $m_i(t)$ of the type i equipment elements, $i=1,\ldots,k_i^{\max}$, that will be acquired in year t, $t\in[t_0,T]$. Combining together all these quantities for each period $k=k_i^{\max}$ in year t and selecting the maximal quantities for each type basic element, we obtain the predicted quantity of the basic elements of each type that have to be acquired in year t, $t\in[t_0,T]$. Now we can determine the function $a(t)$ and then write off all the basic elements of the production system that entered service before year $a(t)$.

2.3.4 Statement of Optimization Problem

As previously mentioned, the integral model (6.26)-(6.27) usually has more unknowns than equations. Therefore, a unique solution cannot be found and, in order to complete the problem formulation, it is necessary to introduce an optimality criterion. Then the problem in the model becomes an optimization problem on some future planning interval $[t_0,T]$.

Let us take as the optimality criterion *the minimization of operating expenditures* over the entire planning interval $[t_0,T]$ *for implementation of the entire specified schedule of operations*:

$$S=\sum_{i=1}^{n_i}\sum_{j=1}^{N_i}\int_{t_0}^{T}\left[\int_{a_j(t)}^{t}(\sigma_{ij}(\tau,t)+q_{ij}(\tau))m_{ij}(\tau)d\tau+\lambda_j(t)m_{ij}(t)\right]dt. \quad (6.30)$$

where

$\lambda_i(t)$ is the balance value of a type j basic element acquired in year t,

$\sigma_{ij}(\tau,t)$ describes the specific operating costs incurred during the i-th operation in year t of a single assembly with basic elements of type j that was put into operation in year τ (including the cost of maintenance, repair, modernization, etc.),

$q_{ij}(\tau)$ is the specific unit cost of labor for implementation of the i-th operation of a single assembly with a basic element of type j,

S is the total operating costs incurred in all operations over the entire interval $[t_0,T]$.

Having the objective function (6.30), we can formulate the problem of optimal replacement of the equipment elements in a production system as an optimal control problem.

The optimization problem (OP) is to determine the sought-for functions $m_{ij}(t)>0$ and $a_j(t)>0$, $i=1,\dots,n_k$, $j=1,\dots,N_i$, $t\in[t_0,T)$, that minimize the functional S:

$$S \longrightarrow \min \quad (6.31)$$

under the following equality constraints:

$$b_i(t) = \sum_{j=1}^{N_i} \int_{a_j(t)}^{t} \beta_{ij}(\tau,t) m_{ij}(\tau) d\tau, \quad i = 1,\ldots,n_k, \tag{6.32}$$

the inequality constraints:

$$m_{ij}(t) \geq 0, \qquad a_j(t) < t, \tag{6.33}$$

$$i=1,\ldots,n_k,\ j=1,\ldots,N_i,\ t \in [t_0,T],$$

and the initial conditions:

$$a_j(t_0) = a_j^0,\ m_{ij}(\tau) = m_{ij}^0(\tau), \tag{6.34}$$

$$i=1,\ldots,n_k,\ j=1,\ldots,N_i,\ \tau \in [\tau_0,\ t_0].$$

Since the current labor expenditures $q_{ij}(\tau)m_{ij}(\tau)$ are already included in the total cost S of production, the manpower demand given in (6.28) is not considered separately in the statement of the optimization problem.

In this problem, the functions $\sigma_{ij}(\tau,t)$, $\lambda_{ij}(t)$, $q_{ij}(\tau)$, $\beta_{ij}(\tau,t)$, and $b_i(\tau)$ are assumed to be known. Solving the optimization problem (6.31)-(6.34) provides us with the set of functions $m^*_{ij}(t)$, $a^*_j(t)>0$, $i=1,\ldots,n_k$, $j=1,\ldots,N_i$, $t \in [t_0,T)$. Once the functions m^*_{ij} have been determined, it is possible to find the required number of new basic equipment elements of type j:

$$m_j(t) = \sum_{i=1}^{n_k} m_{ij}(t), \qquad i = 1,\ldots,N_i. \tag{6.35}$$

Once the functions a^*_j, $j=1,\ldots,N_i$, have been determined, we obtain a strategy for removing from service in year t all the basic elements of type j that were put into operation before the year $a^*_j(t)$. The optimal useful life (service time) of the basic elements of type j is equal to $L_j(t)=t-a^*_j(t)$.

3. QUALITATIVE ANALYSIS OF OPTIMAL EQUIPMENT REPLACEMENT

This section examines the qualitative behaviour of some optimization problems (OP) formulated above. Under simplified assumptions, the analytical study of the problems leads to non-trivial results, in particular, to so-called *turnpike properties* of the optimal scrapping time for the obsolete equipment. The analysis consists of finding "efficient" trajectories (turnpikes) which are close in some sense to a solution of the optimization problem and have a simpler structure. The investigation of turnpike properties is helpful when the optimization problem cannot be solved directly. Generally speaking, turnpike theorems indicate certain basic patterns of economic development. The analysis provided below provides an insight into the rational scrapping and replacement of obsolete industrial equipment in production systems.

Section 3.1 describes a general investigative technique used for the optimization problems in integral models. Section 3.2 investigated the aggregated optimization problem, derives an integral-functional equation for turnpike trajectories and establishes the turnpike properties of optimal trajectories (the optimal equipment useful life) in the cases of infinite- and finite-horizon optimization. Section 3.3 is devoted to optimization of equipment replacement in disaggregated integral models of production systems. Section 3.4 discusses the results as well as certain residual problems. Detailed mathematical aspects of the results are provided in Section 4.

3.1 About Optimal Control Problems in Integral Models

The problems considered here – the optimal renovation of equipment in production systems – belong to optimal control problems for the dynamic systems governed by specific integral equations. While this field of optimal control is quite complicated, it is not new. Investigation techniques for such problems fall into the optimization methods in abstract (functional) spaces. These methods have been the subject of intensive study during last forty years (Pontryagin et al, 1962; Rocafeller, 1970; Pshenichniy, 1971; Warga, 1972; Neustadt, 1976; Ioffe & Tichomirov, 1979; Vasiliev, 1981; Alekseev et al, 1987; Rubinov, 2000; and others).

Specific optimal control problems for dynamic systems governed by integral equations have been studied in (Vinokurov, 1969; Angell, 1976; Bakke, 1974; Carlson, 1987; Medhin, 1986; Shaikhet, 1985; Leugering, 1987; Connor, M.A. and Hood, 1973; etc.) where the variation techniques were developed and the maximum principle was established for such systems.

A major new feature of the control problems under study is the *new type of sought-for control functions* that are the limits of integration in integral equations. Additional complications in various problems of technological development may also include :

- differential (non-holonomic) restrictions on sought-for variables,
- mixed restrictions on the optimal controls and phase variables;

A technique for treatment of the above-mentioned problems was developed in the series of publications (Glushkov, Ivanov & Yatsenko, 1980, 1982; Yatsenko, 1991; Hritonenko & Yatsenko, 1994; Hritonenko & Yatsenko, 1996; etc.). A brief overview of this technique is provided below. We concentrate on its similarity and differences from other optimal control problems for integral equations. More detailed proofs can be found in (Hritonenko & Yatsenko, 1996).

3.1.1 General Statement of Optimal Control Problem

Let us consider the following general statement of the optimal control problem for Volterra integral equations with *sought-for control functions* in the limits of integration. It consists of the minimization of a functional

$$I = \int_{t_0}^{T} \Phi(\mathbf{x}, \mathbf{y}, \mathbf{z}, t)dt \rightarrow \min_{\mathbf{x},\mathbf{y},\mathbf{z}} \tag{6.36}$$

under the equality constraints (model equations or *state equations*):

$$x_i(t) = \sum_{j=1}^{n} \int_{z_j(t)}^{t} K_{ij}(\tau,t) y_i(\tau,t) x_j(\tau) d\tau + f_i(t), \; i = 1,\dots,n+m, \tag{6.37}$$

under inequality constraints of the type:

$$\sum_{i=1}^{n} y_i(t) \le 1, \quad x_i(t) \ge 0, \quad z_j'(t) \ge 0, \quad z_j(t) < t. \tag{6.38}$$

and the initial conditions on the prehistory of the process:

$$z_j(t_0)=z_j^0,\ \ x_j(\tau)=x_j^0(\tau),\ \ \tau\in[\,z_j(t_0),\,t_0],\ \ j=1,\ldots,n. \qquad (6.39)$$

As the sought-for functions **x, y, z** (unknowns of the problem), we will consider *a subset* of the functions $x_j(t)$, $y_i(t)$, $z_j(t)$, $t\in[t_0, T)$, i.e.:

$$\begin{aligned}\mathbf{x} &= \{x_j(.)\},\ \ j\in J_x\subset\{1,\ldots,n+m\},\\ \mathbf{y} &= \{y_i(.)\},\ \ i\in J_y\subset\{1,\ldots,n\},\\ \mathbf{z} &= \{z_j(.)\},\ \ j\in J_z\subset\{1,\ldots,n\}.\end{aligned}$$

As in any optimal control problem, we assume that the total number of sought-for functions is greater than the number $n+m$ of the state equations (6.37). Using the equations (6.37), we can often exclude some of the unknowns from the problem and consider them as *dependent (phase) variables*. Then the rest of the unknowns are the *independent control variables*. The choice of the phase variables is not trivial and depends on many factors. We will use different choices in the problems below.

The smoothness requirements for the sought-for variables depend on the specifics of a problem. Usually, the functions $y_j(.)$ are assumed to be measurable, and the functions $x_j(.)$ and $z_j(.)$ are continuous almost everywhere (a.e.).

Let us define **U** as a set of the control variables **x, y, z** that satisfy the constraints (6.38). A solution (if it exists) of the problem is denoted as $(\mathbf{x}^*,\mathbf{y}^*,\mathbf{z}^*)$.

3.1.2 Necessary Conditions of Extremum

The investigation methods for such problems are based on variation techniques. Let us give some small variations (increments) $\delta\mathbf{x}$, $\delta\mathbf{y}$, $\delta\mathbf{z}$ to the sought-for variables **x, y, z**, i.e. consider the functions $\mathbf{x}(t)+\delta\mathbf{x}(t)$, $\mathbf{y}(t)+\delta\mathbf{y}(t)$, $\mathbf{z}(t)+\delta\mathbf{z}(t)$, $t\in[t_0,T)$. Then the functional I gets some increment

$$\delta I = I\,(\mathbf{x}+\delta\mathbf{x},\ \mathbf{y}+\delta\mathbf{y},\ \mathbf{z}+\delta\mathbf{z}) - I\,(\mathbf{x},\ \mathbf{y},\ \mathbf{z})$$

caused by the increments $\delta\mathbf{x}$, $\delta\mathbf{y}$, $\delta\mathbf{z}$.

Since the domain **U** is closed, a *general necessary condition of extremum* consists of the non-negativity of the variation of the functional (6.36):

$$\delta I\,(\mathbf{x}, \mathbf{y}, \mathbf{z}, \delta\mathbf{x}, \delta\mathbf{y}, \delta\mathbf{z}) \geq 0 \tag{6.40}$$

for all admissible variations $\delta\mathbf{x}$, $\delta\mathbf{y}$, $\delta\mathbf{z}$ (Ioffe & Tichomirov, 1979; Vasiliev, 1981). The admissible variations $\delta\mathbf{x}(t)$, $\delta\mathbf{y}(t)$, $\delta\mathbf{z}(t)$, $t\in[t_0,T)$, are any variations such that $\mathbf{x}(t)+\delta\mathbf{x}(t)$, $\mathbf{y}(t)+\delta\mathbf{y}(t)$, and $\mathbf{z}(t)+\delta\mathbf{z}(t)$ do not violate the conditions (6.37)-(6.39).

If functional (6.36) is differentiable, than we can separate the linear part of the increment δI and represent the condition (6.40) as the following:

$$\begin{aligned}\delta I &= \langle \partial I/\partial\mathbf{x}, \delta\mathbf{x}\rangle + \langle \partial I/\partial\mathbf{y}, \delta\mathbf{y}\rangle + \langle \partial I/\partial\mathbf{x}, \delta\mathbf{z}\rangle + o(\delta\mathbf{x}, \delta\mathbf{y}, \delta\mathbf{z}) = \\ &= \int_{t_0}^{T} [\sum_j \partial\Phi(t)/\partial x_j \cdot \delta x_j(t) + \sum_j \partial\Phi(t)/\partial y_i \cdot \delta y_i(t) + \\ &\quad + \sum_j \partial\Phi(t)/\partial z_j \cdot \delta z_j(t)]dt + o(\delta\mathbf{x}, \delta\mathbf{y}, \delta\mathbf{z}) \geq 0. \end{aligned} \tag{6.41}$$

Here $\langle x, u\rangle$ denotes the value of a linear functional $c\in B^*$ at the element $u\in B$, where B^* is the space dual to B. The expression $\langle x, u\rangle$ is the scalar product if B is a Hilbert space. The functions $\partial\Phi/\partial x_j$, $\partial\Phi/\partial y_i$, $\partial\Phi/\partial z_j$ are called the *functional derivatives* of the functional I (with respect to corresponding functions x_j, y_i, z_j).

From the viewpoint of abstract optimization theory (Rocafeller, 1970; Pshenichniy, 1971; Neustadt, 1976; Ioffe & Tichomirov, 1979), the functional I is an abstract function of the variable $(\mathbf{x}, \mathbf{y}, \mathbf{z})$ that belongs to an abstract functional space. Formula (6.41) represents the definition of *the Freshet derivative*

$$I' = (\partial I/\partial\mathbf{x}, \partial I/\partial\mathbf{y}, \partial I/\partial\mathbf{z})$$

of the functional I with respect to the variable $(\mathbf{x}, \mathbf{y}, \mathbf{z})$. Using the Freshet derivative I', the necessary condition (6.40) produces the following *necessary condition of extremum* as a corollary (Vasiliev, 1981) :

If $\mathbf{U}^*$ is the set of all solutions of the optimal control problem (6.36)-(6.39), then for any point $(\mathbf{x}^*,\mathbf{y}^*,\mathbf{z}^*)\in\mathbf{U}^*$ the following inequality is true :

$$\langle I'(\mathbf{x}^*,\mathbf{y}^*,\mathbf{z}^*), (\mathbf{x}, \mathbf{y}, \mathbf{z})-(\mathbf{x}^*,\mathbf{y}^*,\mathbf{z}^*) \rangle \geq 0 \qquad (6.42)$$

for all $(\mathbf{x},\mathbf{y},\mathbf{z})\in\mathbf{U}$. If the solution $(\mathbf{x}^*(t), \mathbf{y}^*(t), \mathbf{z}^*(t))$ is internal in the domain $\mathbf{U}^*$ for all $t\in[t_0,T)$, then condition (6.42) is equivalent to the following equality:

$$I'(\mathbf{x}^*, \mathbf{y}^*, \mathbf{z}^*) \equiv 0. \qquad (6.43)$$

To prove that a trajectory $(\mathbf{x},\mathbf{y},\mathbf{z})\in\mathbf{U}$ obtained from the necessary conditions like (6.42) is a solution of the optimization problem under study, we use a standard *sufficient condition of extremum* based on the second variation of the functional I. If the second variation has the same sign in a neighborhood of the trajectory $(\mathbf{x},\mathbf{y},\mathbf{z})$ then the local extremum (minimum or maximum) has been reached (Ioffe & Tichomirov, 1979; Vasiliev, 1981). Investigation of the second variation is sufficiently more difficult than the necessary conditions; therefore, this approach has been fully implemented only for certain basic problems. Such analysis is provided for the aggregated optimization problem (6.51)-(6.56) in Sections 3.2 and 4 below.

In some cases (including the above aggregated model), it is possible to prove the convexity of the optimization problem under study. This extends the qualitative results obtained for local optimal trajectories as properties of the global optimum (Rubinov, 2000).

Further possible development of necessary conditions (6.42) leads to a *maximum principle for integral equations* (6.36). This approach was employed in several papers (Vinokurov, 1969; Angell, 1976; Bakke, 1974; Carlson, 1987) for the standard Volterra integral equations (without sought-for functions in the integration limits). It may be successfully reproduced for the Volterra integral equations (6.37) with sought-for functions in the integration limits. However, this section mainly focuses on internal solutions of the optimal control problems under study. So, our investigation technique is mainly based on the conditions (6.41) and (6.43). The actual form of the functional derivatives in (6.41) for the integral models with controllable memory of type (6.37) is important and reflects the specifics of these problems and their connection with other similar investigations. It is discussed below.

3.1.3 Lagrange Multipliers Method

If the system of state equations (6.37) (an integral dynamical model) is complex, we can use *the Lagrange multipliers method* to exclude the dependent (phase) variables and produce a system of dual equations (adjoint equations) for the optimization problem under study. In this case, the functional I depends only on the independent control functions and the phase variables are defined from the state equations (6.37). It is not necessary in simpler cases, for example, there is no adjoint equation in aggregated optimization problem (6.23) (see section 3.2 below).

We restrict ourselves with demonstration of *Lagrange multipliers method* for a special case of the problem (6.36)-(6.39) when m=0. Then the number of sought-for functions $x_j(.)$ is the same as the number n of the state equations (6.36) and we can consider them as the phase variables. Under the given controls $z_j(.)$, $y_j(.)$, the functions $x_j(.)$ are uniquely determined from the system of linear Volterra equations (6.37) of the second kind.

For simplicity's sake, we also assume that the right end of the trajectory $\mathbf{x}(T)$ is free and the time T (the right end of the planning horizon) is fixed. The left end of the trajectory is fixed by virtue of (6.39).

Let us introduce the following Lagrange multipliers

$$\boldsymbol{\varphi}(t) = (\varphi_1(t), \ \ldots \ , \ \varphi_n(t))$$

for taking the equality constraints (6.37) into account. The Lagrange functional in this case can be constructed as follows (Vasiliev, 1981):

$$L(\mathbf{x},\mathbf{y},\mathbf{z},\varphi) = \varphi_0 I(\mathbf{x},\mathbf{y},\mathbf{z}) + \sum_{i=1}^{n} \int_{t_0}^{T} \varphi_i(t) \times [x_i(t) -$$

$$- \sum_{j=1}^{n} \int_{z_j(t)}^{t} K_{ij}(\tau,t) y_i(\tau) x_j(\tau) d\tau - f_i(t)] dt \qquad (6.44)$$

We give some increments $\delta z_j(t)$, $\delta y_j(t)$, $\delta x_j(t)$ to the sought-for variables z_j, y_j, x_j, $t \in [t_0, T)$, and determine the corresponding increment of the functional (6.44). After the interchange of the order of integration and similar transformations, the linear part of the variation is described by the following formula (Yatsenko, 1991):

$$\delta L = \int_{t_0}^{T} \{ \sum_{j=1}^{n} \left[\varphi_0 \frac{\partial \Phi(t)}{\partial x_j} + \varphi_i(t) - \sum_{j=1}^{n} y_i(\tau) \int_t^{\bar{z}^{-1}{}_j(t)} K_{ij}(t,\tau)\varphi_j(\tau)d\tau \right] \delta x_i(t) +$$

$$+ \sum_{i=1}^{n} \left[\varphi_0 \frac{\partial \Phi(t)}{\partial y_i} - x_i(\tau) \int_t^{\bar{z}^{-1}{}_j(t)} K_{ij}(t,\tau)\varphi_i(\tau)d\tau \right] \delta y_i(t) +$$

$$+ \sum_{j=1}^{n} \left[\varphi_0 \frac{\partial \Phi(t)}{\partial z_j} + \sum_{j=1}^{n} K_{ij}(z_i(t),t) y_i(z_i(t)) x_j(z_i(t)) + \varphi_i(t) \right] \delta z_i(t) \} dt \qquad (6.45)$$

Here $z_j^{-1}(.)$ is the inverse of the function $z_j(.)$ and the function $\bar{z}_j^{-1}(.)$ is the minimum of $z_j^{-1}(.)$ and T:

$$\bar{z}_j^{-1}(t) = \begin{cases} z_j^{-1}(t), & t \in [t_0, z_j(T)], \\ T, & t \in [z_j(T), T), \end{cases} \qquad (6.46)$$

Since $z_j(t) < t$, it is always the case that $t < z_j^{-1}(t)$.

As shown in (Yatsenko, 1991), under natural assumptions about positiveness and smoothness of the given functions $K_{ij}(\tau, t)$, $f_i(t)$, $x_j^0(\tau)$, the state equations (6.36) (the direct problem) have a unique positive solution $x_j(t)$, $j=1,\ldots,n$, $t \in [t_0,T)$, for any given functions $z_j(t)$, $y_j(t)$, $j=1,\ldots,n$, $t \in [t_0,T)$, from the domain (6.38).

Therefore, the conditions $x_i(t) \geq 0$ in (6.38) are never reached and the phase variables $x_i(.)$ belong to an open domain. It follows that the variations $\delta x_j(t)$ may be arbitrary chosen and we have to set the coefficients at the variations $\delta x_j(t)$ in the expression (6.45) equal to 0. It leads to the following *dual system (adjoint equations)* for the optimal control problem:

$$\varphi_i(t) = \sum_{j=1}^{n} y_i(\tau) \int_t^{\bar{z}^{-1}{}_j(t)} K_{ij}(t,\tau)\varphi_j(\tau)d\tau + \varphi_0 \frac{\partial \Phi(t)}{\partial x_j}, \quad i = 1,..n. \qquad (6.47)$$

The system (6.47) of Volterra linear integral equations of the second kind has a unique solution $\varphi_j(.)$, $j=1,\ldots,n$, under the same assumptions as the direct system (6.37) has a solution $x_j(.)$, $j=1,\ldots,n$. It can be observed that, if the direct problem (the state equations (6.37)) includes the integration over some unknown past intervals $[z_j(t), t]$, then the dual problem (adjoint equations (6.47)) includes the integration over unknown future intervals $[t, \bar{z}_j^{-1}(t)] \subset [t, T]$.

Now, the variation of the functional may be represented via the variations $\delta y_j(.)$, $\delta z_j(.)$ of the independent controls only as

$$\delta I = \int_{t_0}^{T} \left[\sum_{j=1}^{n} \frac{\partial I(t)}{\partial x_j} \delta x_i(t) + \sum_{i=1}^{n} \frac{\partial \Phi(t)}{\partial z_i} \delta z_i(t) \right] dt, \tag{6.48}$$

and the set of the functional derivatives

$$I'(\mathbf{y}, \mathbf{z}) = \left\{ \frac{\partial \Phi(t)}{\partial y_j}, \frac{\partial \Phi(t)}{\partial z_j} \right\}_{i=1}^{n}, \tag{6.49}$$

$$\frac{\partial \Phi(t)}{\partial y_j} = \varphi_0 \frac{\partial \Phi(t)}{\partial y_i} - x_i(t) \int_t^{\bar{z}^{-1}_j(t)} K_{ij}(t,\tau)\varphi_i(\tau) d\tau, \tag{6.50}$$

$$\frac{\partial \Phi(t)}{\partial z_j} = \varphi_0 \frac{\partial \Phi(t)}{\partial z_j} + \sum_{j=1}^{n} K_{ij}(z_i(t),t) y_i(z_i(t)) x_j(z_i(t)) + \varphi_i(t).$$

represents *the gradient of the functional I by virtue of the state equations* (6.36) (Rocafeller, 1970; Pshenichniy, 1971; Vasiliev, 1981; etc.).

To calculate the value of the gradient (6.49) under the given controls, it is necessary to determine $\mathbf{x}(.)$ from the direct system (6.37) first and then $\varphi(.)$ from the dual system (6.47).

Remark. Optimal control theory is a generalization of finite-dimensional optimization (mathematical programming). In mathematical programming, the sought-for control variable is an n-dimensional vector $\mathbf{x}=\{x_1,\ldots,x_n\}$ and, correspondingly, the gradient is also an n-dimensional vector grad $f = \{\partial f/\partial x_1, \ldots, \partial f/\partial x_n\}$. In optimal control theory, the sought-for control variable is a function $x(t)$ defined on the interval $[a,b]$, the gradient $I'(t)$ is also a function on $[a,b]$, and the integral over $[a,b]$ in the expression (6.48) plays the role of the sum in the scalar product $\langle x, u\rangle$ in the finite-dimensional space. Similarly as a component $\partial f/\partial x_k$ of the finite-dimensional gradient characterizes the increase of the function f in the direction of the axis x_k, the functional gradient $I'(t_k)$ at a fixed point $t_k \in [a,b]$ characterizes the increase of the functional I at this point t_k.

Other more intricate optimal control problems for the integral equations with controllable memory are investigated in (Yatsenko, 1991).

3.1.4 Novelty and Common Features

While the direct problem (the state equations (6.37)) contains the integrals over some unknown past intervals $[z_j(t), t]$, the expressions (6.50) in the gradient (6.49) and the dual problem (6.47) contain the integrals over some unknown future intervals $[t, z_j^{-1}(t)]$. This feature is typical for all optimal control problems that involve the integral dynamical models with controllable memory (Glushkov, Ivanov & Yatsenko, 1980, 1982; Ivanov & Yatsenko, 1986; Yatsenko, 1991; Hritonenko & Yatsenko, 1996; and others).

The above observation is in complete agreement with other investigations on the optimal control of integral and differential equations, namely:

- As shown in (Vinokurov, 1969; Angell, 1976; Bakke, 1974; Carlson, 1987; Medhin, 1986; Corduneanu, 1991), if the dynamic system is optimized over the horizon $[t_0, T]$ and is governed by *standard Volterra linear integral equations* of the second kind, then its dual problem includes integrals over the future interval $[t, T]$.
- Following (Pontryagin et al, 1962; Warga, 1972; and other), if the dynamic system is optimized over the horizon $[t_0, T]$ and is governed by *ordinary differential equations* with initial conditions at $t=t_0$, then its dual problem is represented by ODEs with initial conditions at $t=T$. There is no direct hereditary effect (after-effect) in the latter case of ODEs, but we still have to solve a so-called *boundary problem of the maximum principle.*

In all cases, the optimal control of dynamic systems (causal equations) leads to solving boundary problems (non-causal equations). The main innovation of the optimal control problems for the integral equations with controllable memory is that the corresponding system of boundary equations still contains the unknown intervals of integration $[z_j(t), t]$ and $[t, z_j^{-1}(t)]$ (inside the total horizon $[t_0, T]$).

3.2 Optimal Equipment Replacement in Aggregate Model

In accordance with Section 2.3, our aggregated optimization problem (OP) consists of the determination of the functions $m(t)$, $a(t)$, $t\in[t_0, T)$, $T\leq\infty$, which maximize the following objective function:

$$I = \int_{t_0}^{T} \rho(t)[\int_{a(t)}^{t} l(\tau)m(\tau)d\tau + P(t)m(t)]dt \xrightarrow[a,m]{} \min, \qquad (6.51)$$

under the equality constraint:

$$K(t) = \int_{a(t)}^{t} m(\tau)d\tau, \qquad (6.52)$$

the inequality constraints:

$$0 \leq m(t) \leq M(t), \qquad (6.53)$$

$$a'(t) \geq 0, \quad a(t) < t, \qquad (6.54)$$

and the initial conditions:

$$a(t_0) = a_0 < t_0, \qquad m(\tau) = m_0(\tau), \quad \tau\in[a_0, t_0]. \qquad (6.55)$$

From the economic viewpoint, this OP is a classic problem of expenditure minimization under the given production output in the non-standard economic-mathematical model (6.51)-(6.52). This model takes *embodied technological progress* into account as decrease of the function $l(.)$ (it means that newer equipment is more efficient). The control influence in the OP is also non-classic and is made by the change of the renovation intensity of the equipment technological structure. As a part of this non-standard control type, the useful life of the equipment is considered as a sought-for control function that brings a radically new feature in comparison with other similar investigations.

From the mathematical point of view, this optimization problem is analogous to those studied in (Yatsenko, 1991; Hritonenko and Yatsenko, 1996). The presence of sought-for functions in the limits of integration generates some difficulties for qualitative research. The method developed in (Hritonenko and Yatsenko, 1996) makes it possible to obtain essential results in the study of optimal trajectories, such as an explicit form solution of an optimization problem in particular cases and different turnpike properties.

3.2.1 Structure of Aggregated Optimization Problem

We assume that the given functions $\rho(.)$, $l(.)$, $K(.)$, $M(.)$ are Lipschitz continuous, $m_0(.)$ is piece-wise continuous, all these functions are positive and satisfy (6.52)-(6.55) at $t = t_0$.

Let $m(t)$, $t\in[t_0, T)$, be an *independent control variable* of OP (6.51)-(6.55). Then the function $a(t)$, $t\in[t_0, T)$, is a dependent (*phase*) sought-for variable. We define **U** as a set of the measurable control variables $m(.)$ that satisfy the condition (6.53). Let (m^*, a^*) be a solution (if it exists) of the OP.

Severe complications are caused in the OP by the phase differential constraint (6.54). A possible technique for such OP analysis was mentioned above in Section 3.1 and consists of the reduction to a standard OP by narrowing of the domain **U** of the control $m(.)$. In accordance with this technique, we replace the constraint (6.54) for the phase variable $a(.)$ with the following more strict domain $\mathbf{U}_1$ for the control $m(.)$:

$$m_{\min}(t) \le m(t) \le M(t), \tag{6.56}$$

$$\text{where} \quad m_{\min}(t) = \max\{0, K'(t)\}.$$

From the economic point of view, the regime $m \equiv m_{\min}$ of minimum possible renovation of the firm equipment is trivial. Then $a'(t) = 0$ or $m(t) = 0$ for $t\in[t_0, T)$ in view of (6.56) and (6.52). As will be seen later, the optimal control $m^*(.)$ (if it exists) does not coincide with the minimum possible one $m_{\min}(.)$ under the *presence of the embodied technological progress* (i.e., if the function $l(.)$ decreases). We will suppose that $l(.)$ is, at least, non-increasing, i.e. the technological progress is continuous.

For any measurable control $m(t)$ that satisfies (6.53) almost everywhere (*a.e.*) for $t\in[t_0,T)$, a unique a.e. continuous function $a(t)$, $t\in[t_0,T)$, exists and satisfies (6.52), (6.54) and a.e. has $a'(t)\ge 0$. The proof of this and certain following mathematical statements can be found in (Hritonenko and Yatsenko, 1996).

Thus, restriction (6.56) is more rigid than (6.53)-(6.54) (the corresponding control set $\mathbf{U}_1 \subset \mathbf{U}$). Let us replace OP (6.51)-(6.55) with OP (6.51)-(6.56) with the narrower domain $\mathbf{U}_1$ of the admissible controls, for which the phase constraint (6.54) is satisfied automatically. OP (6.51)-(6.56) is considered below.

Equation (6.52) can be solved with respect to the phase variable $a(.)$ for any given control $m(.) \in \mathbf{U}_1$. It permits derivation of the gradient of the functional (6.51) (Lemma 1 of Section 4) and the necessary conditions of extremum (Lemma 3 in Section 4).

In accordance with Section 3.1, if the functional $I(m)$ in OP (6.51)-(6.56) is differentiable, then the formal definition of the gradient $I'(t)$, $t \in [t_0, T)$, of the functional $I(m)$ is given by the following formula for the increment δI of functional $I(m)$:

$$\delta I = I(m + \delta m) - I(m) = \int_{t_0}^{T} I'(t)\delta m(t)dt + O(\| \delta m \|^2).$$

It follows from Lemma 1 that, if $m(t) > 0$ for $t \in [t_0, T)$, then the gradient $I'(.)$ of functional (6.51) in OP (6.51)-(6.56) exists and has the following form:

$$I'(t) = \int_t^{\alpha^{-1}(t)} \rho(\tau)[l(t) - l(a(\tau))]d\tau + P(t)\rho(t), \quad t \in [t_0, T), \tag{6.57}$$

where

$$\alpha^{-1}(t) = \begin{cases} a^{-1}(t), & t \in [t_0, a(T)], \\ T, & t \in [a(T), T), \end{cases}$$

and $a^{-1}(.)$ is the inverse function of $a(.)$.

It should be noted that, if the condition $m(t)>0$, $t \in [t_0, T)$, is not valid, then the function $I(m)$ is not differentiable. However, this case is natural from the economic viewpoint and is also considered below.

All mentioned results for the integral macroeconomic models were obtained when local optimization was considered. Here, we prove the convexity of the optimization problem under study. It extends the qualitative results (including the turnpike properties) obtained for local optimal trajectories as properties of the global optimum. The convexity of OP (6.51)-(6.56) is established in Lemma 2 of Section 4. It means that the OP solution (m^*, a^*) is unique (if it exists) and therefore delivers the global optimum of the OP (Rubinov, 2000). Thus, the qualitative properties of optimal trajectories established below are global.

Next, the structure and asymptotics of the solution (m^*, a^*) are studied. The asymptotic analysis is based on the equation $I'(t)=0$, $t\in[t_0, T)$, which is a special non-Volterra integral-functional equation with respect to $a(.)$. The analysis makes it possible to prove important turnpike properties of the optimal trajectory a^*. For more detailed investigation of the solution structure, the cases $T=\infty$ and $T<\infty$ are analyzed separately.

3.2.2 Equation for Turnpike Regimes of Equipment Replacement

The integral-functional equation $I'(t)=0$, $t\in[t_0,\infty)$, or

$$\int_t^{a^{-1}(t)} \rho(\tau)[l(a(\tau)) - l(t)]d\tau = P(t)\rho(t), \quad t\in[t_0,\infty), \tag{6.58}$$

with respect to the sought-for function $a(t)$, $t\in[t_0,\infty)$, plays an important role in qualitative analysis of the OP trajectories. The solution $a(t)$ of equation (6.58) on the infinite interval $[t_0,\infty)$, if it exists, will be denoted as $a^{\sim}(.)$ and called *the turnpike trajectory* of OP (6.51)-(6.56). In accordance with economic content of equation (6.58), we will consider only its monotonic solutions $a(t)$ such that $a(t) <t$.

Equation (6.58) has a clear *engineering interpretation*:
In a rational strategy of equipment renovation under technological change, the profit of putting a new equipment unit into service and removing an older obsolete equipment unit from service should be equal to the price of the new equipment unit. Technological change implies that the newer equipment is more efficient.

The integral-functional equation (6.58) was earlier investigated in (Hritonenko and Yatsenko, 1994; Hritonenko and Yatsenko, 1996). This equation generates a set of solutions $a_T(t)$ for a finite interval $[t_0,T]$. The given functions $l(.)$,$P(.)$, $\rho(.)$ need to satisfy strict conditions for existence of the solutions $a_T(t)$ on large intervals $[t_0,T]$, $T >> t_0$. The problem of the existence and uniqueness of a solution $a^{\sim}(t)$ on the infinite interval $[t_0,\infty)$, is complicated and was resolved for specific cases only, namely, for the exponent, power, and logarithmic functions $l(.)$.

In this section we restrict ourselves to the following case:

$$\begin{aligned} l(t) &= l_0 \exp(-c_1 t), \\ P(t) &= P_0 \exp(-c_2 t), \\ \rho(t) &= \rho_0 \exp(-c_3 t), \\ c_1, l_0, P_0 > 0, \quad & c_2, c_3 \geq 0, \quad 0 < \rho_0 \leq 1. \end{aligned} \tag{6.59}$$

In the case $c_2 > 0$ equation (6.58) has a unique solution $a^{\sim}(t)$, $da^{\sim}/dt>0$, $t\in[t_0,\infty)$, such that:

- if $c_1 < c_2$, then $t-a^{\sim}(t) \to 0$ at $t\to\infty$;
- if $c_1 > c_2$, then $t-a^{\sim}(t) \to \infty$ and $a^{\sim}(t) \to \infty$ at $t\to\infty$;
- if $c_1 = c_2$, then $a^{\sim}(t) \equiv t-L$, $t\in[t_0,\infty)$, where the constant L is defined from the following non-linear equation:

  – at $c_3 > 0$

  $$c_3 \exp(c_1 L) + c_1 \exp(-c_3 L) = (c_3+c_1)(1+P_0/l_0);$$

  – at $c_3 = 0$

  $$(\exp(c_1 L) - 1)/c_1 - L = P_0/l_0 .$$

In particular, $L \approx [2P_0(c_3+c_1)/l_0/c_1/(c_1-c_3)]^{1/2}$ for $0 \leq c_3 < c_1 << 1$.

In the above-mentioned case $c_2 > 0$, equation (6.58) has also some set of solutions $a_T(t)$, $da_T/dt>0$, for any finite interval $t\in[t_0,T]$ and all solutions $a_T(.)$ approach the unique solution $a^{\sim}(.)$ at $T\to\infty$.

In the case $c_2 = 0$, equation (6.58) has a bounded solution $a^{\sim}(.)$ on the infinite interval $[t_0,\infty)$ such that $a^{\sim}(t) \to -\ln(c_3 P_0/l_0)/c_1$ at $t\to\infty$.

In the case $c_2 < 0$, equation (6.58) has no solution for any significantly large interval $[t_0,T]$ and, naturally, no solution $a^{\sim}(.)$ for the infinite interval $[t_0,\infty)$.

3.2.3 Infinite-Horizon Discounted Optimization

The OP is meaningful at $T = \infty$ (i.e., the improper integral converges in (6.51)) under the following conditions:

$$\int_0^\infty \rho(t)l(t)K(t)dt < \infty, \qquad \int_0^\infty \rho(t)P(t)dt < \infty.$$

Statement 1. The structure of the OP solution.

Let $T = \infty$, the function $l(t)$ strictly decrease and a function $a^{\sim}(t)$, $t\in[t_0,\infty)$, exist such that $I'(t) \equiv 0$ for $a(.) \equiv a^{\sim}(.)$ and the corresponding $m^{\sim}(.)$ satisfies (6.56).

Then OP (6.51)-(6.56) has a unique solution (m^*, a^*) of the following form:

$$m^*(t) = \begin{cases} m_{\min}(t) \text{ or } M(t), & t \in [t_0, \mu), \\ m^{\sim}(t), & t \in [\mu, \infty), \end{cases} \tag{6.60}$$

$$a^*(t) = \begin{cases} a_{\text{bnd}}(t), & t \in [t_0, \mu), \\ a^{\sim}(t), & t \in [\mu, \infty), \end{cases}$$

where the function

$$a_{\lim}(t) = \begin{cases} a_{\min}(t) & \text{if} \quad a_0 > a^{\sim}(t_0), \\ a_{\max}(t) & \text{if} \quad a_0 < a^{\sim}(t_0), \end{cases}$$

is found from (6.52) at $m^*\equiv m_{\min}$ or $m^*\equiv M$, the function $m^{\sim}$ is found from (6.52) at $a\equiv a^{\sim}$, and the instant μ is determined from the condition $a_{\text{bnd}}(\mu)=a^{\sim}(\mu)$.

The proof of this statement includes the following steps. First, it may be directly checked that the pair of functions (m^*, a^*) is admissible in the set

U. Next, using the necessary and sufficient extremum conditions one can prove that the functions (m^*, a^*) represent an OP solution. The complete proof of this statement is given in Section 4 of this Chapter.

Statement 1 and the properties of equation (6.58) described in subsection 3.2.2 indicate that OP (6.51)-(6.56) may have turnpike properties in the case $T=\infty$.

Statement 2. Turnpike theorem in the strongest form.

If $T=\infty$, $l(t)=l_0\exp(-c_1 t)$, $P(t)=P_0\exp(-c_2 t)$, $\rho(t)=\rho_0\exp(-c_3 t)$,

$$c_1, c_2, l_0, P_0, \rho_0 > 0, \quad c_3 \geq 0,$$

then OP (6.51)-(6.56) has a unique turnpike trajectory $a\tilde{}(t)$, $da\tilde{}/dt>0$, $t\in[t_0,\infty)$, that, in general, does not satisfy the initial condition (6.55).

At $c_1<c_2$ this function $a\tilde{}(t)\to t$ for $t\to\infty$, at $c_1>c_2$ $t-a\tilde{}(t)\to\infty$ and $a\tilde{}(t)\to\infty$ for $t\to\infty$, and $a\tilde{}(t)\equiv t-L$ at $c_1=c_2$ (the constant L is defined above).

The unique solution a^* of OP (6.51)-(6.56) at $T=\infty$ coincides with $a\tilde{}(.)$ starting with an instant $\mu\geq t_0$: $a^*(t)\equiv a\tilde{}(t)$, $t\in[\mu,\infty)$, provided the corresponding function a^* satisfies (6.56).

The value $\mu-t_0$ is defined only by the value of the deviation $|a\tilde{}(t_0)-a_0|$ ($\mu=t_0$ if $a\tilde{}(t_0)=a_0$).

The Proof: existence of the unique function $a\tilde{}(t)$, $da\tilde{}/dt>0$, $t\in[t_0,\infty)$, has been shown in Section 3.2.2 and the existence of the unique solution $a^*(.)$ follows from the Statement 1.

3.2.4 Finite-Horizon Optimization

At $T<\infty$ the structure of the OP solutions is more complicated than the $T=\infty$ case. In particular, the following well-known result about investment inefficiency near the end of the planning horizon is valid for the OP:

Statement 3. No investment near the end of planning horizon.

There is an instant Θ, $t_0\leq\Theta<T$, such that the optimal investment $m^*(t)$ is minimally possible at the end (Θ, T) of the planning horizon $[t_0, T]$: $m^*(t)\equiv m_{\min}(t)$ for $t\in(\Theta, T)$.

The proof follows from the analysis of the expression (6.57) for the gradient $I'(t)$ of the OP at $t\in\to T$ (then $I'(t)>0$).

When the asymptotic behavior of the OP is studied ($T\to\infty$), the structural properties of the OP solution $a^*(.)$ allow us to prove a turnpike property that is weaker than that for the case $T=\infty$.

Statement 4. Turnpike theorem in normal form.

Let:

1. the given function $l(t)$ be strictly decreasing, $l'(t)<0$ be non-decreasing, $t\in[t_0, T]$;
2. the given function $M(t)$ be bounded at $t\to\infty$;
3. the set of the solutions $a_T(t)$, $t\in[t_0, T]$, of equation (6.58) approach a unique trajectory $a^{\sim}(t)$, $t\in[t_0,\infty)$, as $T\to\infty$, such that $da^{\sim}/dt>0$ and $a^{\sim}(t)\to\infty$ at $t\to\infty$;
4. the function $m^{\sim}(t)$ defined by $a^{\sim}(t)$ in accordance with equation (6.52) satisfy (6.56);
5. $K'(t)>0$, $K(t)\to\infty$ as $t\to\infty$ or $a^{\sim}(t_0)>a_0$, $m_0(\tau)\geq d_0>0$, $\tau\in[\tau_0, t_0]$.

Then, when $T\to\infty$, the solution $a^*(.)$ of OP (6.51)-(6.56) tends to $a^{\sim}(.)$ on an asymptotically largest part of the interval $[t_0, T]$,

e.g., for any $\varepsilon>0$ the time T_0 exists such that for any $T\geq T_0$ the condition $|a^*(t)-a^{\sim}(t)|<\varepsilon$ is true on some subset $\Delta\subset[t_0, T]$: $\mathrm{mes}(\Delta)/(T-t_0)\to 1$ for $T\to\infty$.

The proof is based on the analysis of the second variation of the functional I and uses a special admissible solution which coincides with $a^{\sim}(.)$ for the asymptotically largest part of the interval $[t_0, T]$. The detailed proof of Statement 4 is given in Section 4.

The third condition of Statement 4 is the most complicated to verify. As has been shown above in section 3.2.2, this condition is valid, at least, in the case (6.59) for $c_2>0$.

The fifth condition of Statement 4 is essential. If it is not satisfied, then the abstract function $I(m)$ is not differentiable, the technique used for the proof of Statement 4 fails and a counterexample can be constructed.

However, as examples imply, Statement 4 is also valid for some cases where its fifth condition does not hold. The general case requires more refined analysis and can be studied on the basis of the notion of the OP *quasi-solution* $a_q(t)$, $t\in[t_0, T]$, that satisfies a necessary and sufficient extremum condition and is not concerned with the initial condition (6.55)

$a(t_0)=a_0$ (Yatsenko, 1991). If such a quasi-solution exists, then the optimal trajectory $a^*(.)$ coincides with it except for an initial finite interval $[t_0, \mu)$.

In some cases when the quasi-solution is defined explicitly, Statement 3 may be strengthened. For instance, the *strong-form turnpike theorem* was proven in (Hritonenko and Yatsenko, 1996) for a similar optimization problem in an integral model. Such theorems state that under certain conditions the optimal trajectory $a^*(.)$ could sufficiently deviate from a turnpike trajectory $a^*(.)$ only at the beginning and at the end of the planning interval $[t_0, T]$.

3.2.5 Discussion of Results

Thus, a turnpike property in the strongest form has been proven for the infinite-horizon problem of the optimization of equipment useful life. This property states that the optimal equipment scrapping date $a^*(.)$ coincides with a turnpike trajectory $a^{\sim}(.)$ except for some initial finite interval.

However, for the analogous finite-horizon optimization problem we can show only a weaker turnpike property: $a^*(.)\rightarrow a^{\sim}(.)$ as $T\rightarrow\infty$ on an asymptotically largest part of the finite horizon interval $[t_0, T]$ and the optimal trajectories $a^*(.)$, $m^*(.)$ have more complex structure.

This difference between infinite-horizon and finite-horizon problems can be easily explained from an economic viewpoint. Namely, the finite-horizon optimization is less natural than the discounted infinite-horizon optimization because in the latter case all times $t>t_0$ produce equal contribution into an objective functional. It is not so in the case $T<\infty$, and the determination of the end T of the optimization horizon is a well known problem in the economic growth theory.

Applied importance of the established results consists of finding some "efficient" trajectory (the turnpike) close to the optimal equipment useful life but with simpler structure. For many real production systems, it is easier to find a turnpike trajectory and to investigate its properties rather than to solve an optimization problem directly.

The turnpike equipment useful life appears to be independent of the given capital stock $K(t)$ and the initial equipment age structure (the given investments $m_0(\tau)$ on the pre-history $[a(t_0), t_0]$). This optimal strategy is defined only by the historical labour/output coefficient $l(.)$ and the equipment unit cost $P(.)$ that represent the dynamics of the technological progress. So, the turnpike regimes indicate some basic patterns of equipment renovation.

The following properties of the rational (turnpike) useful life $L(t)= t-a^{\sim}(t)$ of the equipment can be figured out (at least, under exponential dynamics of the technological progress and the equipment cost):

- If the relative rate l'/l of the labour/output coefficient is less that the relative rate P'/P of the equipment cost (per efficiency unit), then the equipment useful life $L(t)$ decreases, and the converse.
- If these rates l'/l and P'/P are equal (i.e., the equipment cost/capacity ratio is constant), then the useful life $L(t)$ is constant: $L(t) \equiv L =$ const. The constant L depends on this rate, the equipment cost/capacity ratio, and the discount rate only (the value of L does not depend on the absolute values of the cost $P(t)$ and the labour/output ratio $l(t)$).
- If the equipment cost $P(t)$ per efficiency unit increases (i.e., the capital/labour ratio increases more quickly than the output/labour coefficient), then a turnpike regime of the equipment useful life is impossible.
- The case of a constant equipment price $P(t)$ is critical for the existence of turnpike useful life regimes. In this case, a bounded solution $a^{\sim}$ of the turnpike equation may still exist, so a turnpike property can be valid in the case of infinite-horizon optimization, but it is not true for a finite-horizon case (see the third condition of Statement 3).

The above properties can be used as a basis for decision-making in rational replacement of obsolete industrial equipment.

In (Malcomson, 1975), an integral-functional equation for the optimal useful life of equipment similar to (6.58) was derived and its economic analysis was provided. In particular, the expressions for the implicit rental value of capital equipment and for optimal scrapping criteria were derived. It was shown that, under simplifying assumptions, this equation produces simpler optimal replacement formulas stated in earlier works of G.Terborgh, V.Smith, and H.Brems. It seems likely that J.M.Malcomson derived a dual problem (the adjoint equation) for optimization of equipment useful life in the models with embodied TC for the first time in mathematical economics.

No analytical investigation of the mentioned equation for the optimal equipment useful life was performed in (Malcomson, 1975), and no asymptotic properties of the optimal useful life were found in an analytical way, although some convergence of the upper and lower bounds for the optimal useful life was suggested on the basis of numerical experiments. As noted in (Malcomson, 1975), "this convergence has not yet been confirmed or denied analytically for the models concerned". The turnpike properties established in (Hritonenko and Yatsenko, 1994, 1996) and the present book may be considered as an analytical proof of this convergence.

Some other important implications followed from the analysis provided in the Malcomson model have also been confirmed over the present investigation:

- Scrapping and replacement decisions are independent of product market conditions (the optimal useful life of the equipment does not depend on any of the variables that determine the scale of operations in any material way).
- The equation for determining the optimal useful life is generalized to many types of capital goods for which an independent operating cost function can be determined. It can be shown that the turnpike properties of the OP remain true if we consider other expenditures such as raw material or energy. In doing so, the behavior of optimal trajectories is defined by the most expensive expenditure.

3.3 Optimal Equipment Replacement in Disaggregated Models

This section examines the optimization problem (6.30)-(6.35) for multi-product production system formulated above in Section 2.3.4. As in Section 2.3.3, we restrict ourselves to investigation of the case $N_i=1$, $i=1,\ldots,n_k$, This condition means that the effective set of operated equipment for each i-th operation in the design period k consists of a single type j of basic elements. For simplicity, we also mean $n = n_k$ in this section.

Then the OP is to minimize the functional:

$$S=\sum_{i=1}^{n}\int_{t_0}^{T}\left[\int_{a(t)}^{t}(\sigma_j(\tau,t)+q_i(\tau))m_i(\tau)d\tau+\lambda_i(t)m_i(t)\right]dt\underset{a,m_i}{\rightarrow}\min. \qquad (6.61)$$

by choosing the unknown functions $m_i(t)$ and $a(t)$, $i=1,\ldots,n$, $t\in[t_0, T)$, $T\leq\infty$, under the restrictions:

$$b_i(t)=\int_{a(t)}^{t}\beta_i(\tau,t)m_i(\tau)d\tau,\quad i=1,\ldots,n, \qquad (6.62)$$

$$0 \le m_i(t) \le M_i, \quad i=1,\dots,n, \qquad a'(t) \ge 0, \quad a(t) < t, \qquad (6.63)$$

and the initial conditions:

$$a(t_0) = a^0 \ge 0, \quad m_i(\tau) = m_i^0(\tau), \quad i=1,\dots,n, \quad \tau\in[\tau_0, t_0]. \quad (6.64)$$

The functions $\sigma_i(\tau,t)$, $\lambda_i(t)$, $q_i(\tau)$, $\beta_i(\tau,t)$, and $b_i(\tau)$ are always specified in advance. The given functions $\sigma_i(\tau,t)$, $\lambda_i(t)$, $q_i(\tau)$, $\beta_i(\tau,t)$, $b_i(\tau)$, and $m_i^0(\tau)$ are assumed to be Lipschitz continuous, positive and satisfy (6.62)-(6.64) at $t= t_0$.

The OP (6.61)-(6.64) has only independent control variable, hence, is convenient to choose $a(.)$ as the independent control and $m_i(.)$, $i=1,\dots,n$, as the phase (dependent) variables of the OP. However, in view of the differential constraint $a'(t) \ge 0$, the new independent control

$$v(t) = a'(t) \ge 0, \qquad 0 \le v(t) \le C, \quad C>>1, \quad v\in L_{[t_0,T)}, \qquad (6.65)$$

is introduced, and the equality

$$a(t) = \int_{t_0}^{t} v(\tau)d\tau + a_0, \qquad (6.66)$$

is added to the optimization problem.

Then we have an OP with standard restrictions on the new control variable v. Using the technique described in Section 3.1 and (Hritonenko & Yatsenko, 1996), the following results are established.

Solvability of the state equations (6.62),(6.66). If

$$\partial\beta_i(\tau,t)/\partial\tau>0, \quad \partial\beta_i(\tau,t)/\partial t>0, \quad b_i(t)>0, \qquad (6.67)$$

then for any measurable control v from the domain (6.65) unique measurable functions $m_i(t)$, $i=1,\dots,n$, and an absolutely continuous function $a(t)$, $t\in[t_0, T)$, exist and meet the equations (6.62) and inequalities (6.63) (for sufficiently large constants M_i).

It should be noted that the conditions (6.66) are natural from the economic point of view and indicate the continuity of technological progress, the presence of equipment deterioration, and growth of the enterprise resources. As mentioned in Section 3.1, solvability of the state equations permits to derive the gradient of the OP functional.

The gradient of functional (6.61) in the OP (6.61)-(6.66) has the following form:

$$S'_v(t) = \sum_{i=1}^{n} \int_t^T \left[\varphi_j(\tau)\beta_j(a(\tau),\tau) - (\sigma_j(a(\tau),\tau) + q_i(a(\tau)))\right]\cdot$$

$$\cdot m_i(a(\tau))d\tau. \qquad (6.68)$$

where the sought-for *Lagrange multipliers* $\varphi_i(t)$, $i=1,\ldots,n$, are determined uniquely by the following *dual (adjoint) problem*:

$$\int_t^{\bar{a}^{-1}(t)} \left[\sigma_j(t,\tau) + q_i(t) - \varphi_j(\tau)\beta_j(t,\tau)\right] d\tau + \lambda_j(t) = 0,\ j = 1,\ldots n. \qquad (6.69)$$

$$\bar{a}^{-1}(t) = \begin{cases} a^{-1}(t), & t \in [t_0, a(T)], \\ T, & t \in [a(T), T), \end{cases}$$

where

and $a^{-1}(.)$ is the inverse function of $a(.)$.

The the planning horizon end effect is valid in the case of finite horizon $T<\infty$ (as well as for other OPs in the integral models of technological renovation):

Let v^* be a solution of the optimization problem (6.61)-(6.66). Then there exists an instant θ, $t_0 \le \theta < T$, such that $v^*(t) \equiv 0$ and $da^*/dt \equiv 0$, for $t \in (\theta, T)$. This means that the optimal renovation strategy assumes no renovation at the end of the planning interval $[t_0, T)$.

Turnpike analysis. Following Section 3.1, the qualitative analysis of rational equipment replacement is based on the investigation of internal solutions of the optimization problem if such solutions exist. Then, the equation $S'_v(t) \equiv 0$ produces the following system of $n+1$ *integral-functional equations for turnpike trajectories* $a^{\sim}(t)$, $m^{\sim}_i(t)$, $t \in [t_0, \infty)$ (if such trajectories exist):

$$\sum_{i=1}^{n} \int_t^T \left[\varphi_j(\tau)\beta_j(a(\tau),\tau) - (\sigma_j(a(\tau),\tau) + q_i(a(\tau)))\right]\cdot$$

$$\cdot m_i(a(\tau))d\tau = 0. \qquad (6.70)$$

The integral-functional equations (6.70) should be solved on the infinite interval $[t_0,\infty)$ simultaneously with the dual system (6.68) with respect to the corresponding $\varphi^{\sim}_i(t)$, $i=1,\ldots,n$. Its analysis is very complicated, which is why some partial cases of the equations (6.70) were considered. Here we restrict ourselves with the following case:

Case of equipment with identical technological proportions :

$$\beta_i(\tau,t) = \beta_i(\tau), \qquad \sigma_i(\tau,t) = \sigma_i(\tau),$$

$$\sigma_i(\tau)/\beta_i(\tau) = \sigma_j(\tau)/\beta_j(\tau), \tag{6.71}$$

$$q_i(t)/\beta_i(t) = q_j(t)/\beta_j(t), \quad \lambda_i(t)/\beta_i(t) = \lambda_j(t)/\beta_j(t), \quad i,j=1,\ldots,n.$$

The *interpretation of the conditions* (6.71) consists in the fact that *different operations performed by the equipment units are identical* from the viewpoint of the technological change. Namely, for all different types of operations the ratio of equipment productivities is equal to the ratio of current operating costs, it is also equal to the ratio of labor expenditures and to the ratio of equipment costs.

In this case, the system (6.70) leads to the following one integral-functional equation for a possible turnpike trajectory $a^{\sim}(.)$:

$$\int_t^{a^{-1}(t)} [\sigma_j(t)/\beta_j(t) + q_i(t)/\beta_j(t) - \sigma_j(a(\tau))/\beta_j(a(\tau)) -$$
$$- q_i(a(\tau))/\beta_j(a(\tau))]d\tau + \lambda_j(t)/\beta_j(t) = 0, \; j = 1,\ldots n. \tag{6.72}$$

where

$$\varphi_i(\tau) = [\sigma_i(a(\tau)) + q_i(a(\tau))]\,/\beta_i((a(\tau)), \quad i=1,\ldots,n. \tag{6.73}$$

Thus, we obtain one turnpike equation with respect to the rational (turnpike) equipment useful life $a^{\sim}$, which is analogous to the equation (6.58) in the OP for the aggregated model (see Section 3.2) and possesses interesting general properties.

In particular, if the given functions $\sigma_i(t)$, $\lambda_i(t)$, $q_i(t)$ are constant and the turnpike trajectory $a^{\sim}$ exists, then the turnpike equipment useful life $t-a^{\sim}(t)$ decreases if the relative rate $\beta'_i(t)/\beta_i(t)$ increases, and the converse. The trajectory $a^{\sim}$ possesses some other properties similar to discussed in Section 3.2 for the aggregated model.

3.3.1 Model with Different Lifetimes of Equipment

In a general case, when the assumption (6.71) about identical technological proportions of the equipment fails, there is no common rational (turnpike) equipment useful life $a^{\sim}(.)$ in the disaggregated model (6.62)-(6.64) for the equipment of different types. Relying on the results obtained for various similar integral models in (Yatsenko, 1991; Hritonenko & Yatsenko, 1991), the optimal lifetimes (useful lives) $t-a^{\sim}_i(t)$ of equipment may be expected to be different for various type operations in the case when the different equipment types are not identical from the viewpoint of technological change. For this reason, it is logical to introduce the model analogous to the considered above but with the different scrapping times $a_i(t)$ for various equipment types, $i=1,\ldots,n$.

The corresponding optimization problem consists of determining $2n$ sought-for functions $m_i(t)$ and $a_i(t)$, $i=1,\ldots,n$, $t\in[t_0, T)$, which minimize the functional:

$$S=\sum_{i=1}^{n} \int_{t_0}^{T} \left[\int_{a_i(t)}^{t} (\sigma_j(\tau,t)+q_i(\tau))m_i(\tau)d\tau + \lambda_i(t)m_i(t) \right] dt \to$$

$$\to \min_{a_i, m_i, i=1,\ldots,n} \qquad (6.74)$$

under the restrictions:

$$b_i(t)=\int_{a_i(t)}^{t} \beta_i(\tau,t)m_i(\tau)d\tau, \quad i=1,\ldots,n, \qquad (6.75)$$

$$0 \le m_i(t) \le M_i, \quad a_i'(t) \ge 0, \quad a_i(t) < t, \quad i=1,\ldots,n, \qquad (6.76)$$

and the initial conditions:

$$a_i(t_0) = a^0_i \ge 0, \quad m_i(\tau) = m_i^0(\tau), \quad i=1,\ldots,n, \quad \tau\in[\tau_0, t_0]. \qquad (6.77)$$

The problem (6.74)-(6.77) can be considered as the combination of n simpler optimization problems of the following form:

for fixed i=1,...,n, determine two sought-for functions $m_i(t)$ and $a_i(t)$, $t \in [t_0, T)$, that minimize the functional:

$$S_i = \int_{t_0}^{T} \left[\int_{a_i(t)}^{t} (\sigma_j(\tau,t) + q_i(\tau)) m_i(\tau) d\tau + \lambda_i(t) m_i(t) \right] dt \rightarrow$$

$$\rightarrow \min_{a_i, m_i}. \quad (6.78)$$

under the restrictions (6.75)-(6.77) for the corresponding i. Then

$$S = \sum_{i=1}^{n} S_i, \quad S \underset{a_i, m_i, i=1,\dots,n}{\rightarrow} \min \text{ if } S_i \underset{a_i, m_i}{\rightarrow} \min, \; i = 1, \dots, n. \quad (6.79)$$

At a fixed value of the subscript i, OP (6.78) is similar to optimization problems in aggregated integral model and can be investigated in an analogous way (see Section 3.2). In particular, the integral-functional equation for turnpike trajectory $\tilde{a}_i$ in this OP has a form similar to (6.58); therefore, the optimal trajectories in the composite problem (6.74)-(6.77) for the disaggregated integral model of equipment replacement will possess similar features, including turnpike properties.

In the most general case, we need to consider *a model with equipment interoperability*. Then the analog of the integral equation (6.75) for the equipment demand $b_i(t)$ will contain a sum by several equipment types $m_i(t)$ and a matrix $\{ \beta_{ij}(\tau,t) \}$. The above decomposition of the optimization problem into several simpler problems is no longer possible in the model. While such model provides more flexibility in the management of corresponding production processes, its analysis is significantly more complex and will be considered elsewhere.

3.4 Open Problems

The obtained results represent the first step in modelling of the rational useful life of equipment in the integral economic models with embodied technological change. They are valid for large horizon intervals when the future course of technological change is completely known and the market competition is not critical.

The estimation of a future TC rate represents an important separate problem and has been investigated by many scholars. This rate is influenced by innovation strategy and R&D investment and other factors that require using the models of endogenous TC (Englmann, 1992).

From the viewpoint of the evolutionary self-organizing approach, it is worthwhile to study the collective behavior of several independent production systems (plants) within the framework of a certain integral model of economics (Silverberg, 1987), when some plants use some rational (turnpike) decisions for scrapping obsolete equipment and the rest use other scrapping politics. The goal is to learn whether the plants would have some advantage and to estimate the corresponding time interval. This problem is of interest in the cases when the future TC course is completely known or when it can be approximated only with various stochastic perturbations.

As mentioned in (Bruckner et al., 1993, Silverberg et al., 1988), many phenomena in innovation processes cannot be described as deterministic ones. Construction and investigation of the integral self-organizing models with stochastic effects are among interesting unsolved questions. Such models play a special role in the modelling of market economics. In particular, the self-organizing models of TC must represent the diffusion of new technologies under conditions of market uncertainty as a non-equilibrium stochastic process.

Some interesting conclusions can also be reached by combining the Malcomson model (Malcomson, 1975) and the integral models under study, for example:

- considering the impact of changing product cost on the firm's profit;
- evaluating the rental value of capital equipment and the marginal cost of producing output;
- analyzing different dynamics of optimal equipment useful life produced in various optimization problems as compared with the Malcomson model and the Terborgh formula (Malcomson, 1975).

An important question is to prove that the obtaining qualitative results are global. As shown above, the considered specific OP appears to be convex, but it is not the case for more general integral economic models. So, finding a global optimum and investigating its properties is critical for justification of the obtained qualitative results as global strategies of rational equipment replacement.

The next step would be providing numeric experiments for real economic market data and simulation of rational scrapping of obsolete technologies (equipment). Numeric study of the rational equipment useful life is based on solving turnpike equations similar to (6.58). It was analyzed in (Aistrakhanov and Yatsenko, 1992) where some approximate algorithms were constructed. However, construction of efficient algorithms for finding local and global maximums for the considered class of integral dynamic models represents an open and difficult problem of numeric mathematics.

4. MATHEMATICAL DETAILS AND PROOFS

This section presents detailed proofs of basic statements made in Section 3. Lemmas 1-4 provided at the beginning are necessary to prove the Statements 1-4 of Section 3.

Lemma 1. The differentiability of the functional $I(m)$. If $m(t)>0$, $t\in[t_0, T)$, then the abstract function $I(m)$ in OP (6.51)-(6.56) is differentiable and the increment δI of functional (6.51) is of the form:

$$\delta I = I(m+\delta m) - I(m) = \int_{t_0}^{T} I'(t)\delta m(t)dt + \delta^2 I, \tag{6.80}$$

where the gradient $I'(t)$ of functional (6.51) is given by

$$I'(t) = \int^{\alpha^{-1}(t)} \rho(\tau)[l(t) - l(a(\tau))]d\tau + P(t)\rho(t), \quad t\in[t_0,T), \tag{6.81}$$

the second variation of the functional is

$$\delta^2 I = \int_{t_0}^{T} \rho(t) \int_{a(t)}^{a(t)+\delta a(t)} [l(a(t)) - l(\tau)][m(\tau)+\delta m(\tau)]d\tau dt, \tag{6.82}$$

and the admissible variations $\delta m(t)$, $\delta a(t)$, $t\in[t_0, T)$, of the functions $m(t)$, $a(t)$ must satisfy the equality:

$$\int_{a_{\max}(t)}^{t} \delta m(\tau)d\tau = \int_{a(t)}^{a(t)+\delta a(t)} [m(\tau)+\delta m_{\text{int}}(\tau)]d\tau, \tag{6.83}$$

$$a_{\max}(t) = \max\{a(t), t_0\},$$

$$\delta m_{\text{int}}(\tau) = \begin{cases} \delta m(\tau), & t\in(t_0,T), \\ 0, & t\in[a(t_0),t_0]. \end{cases}$$

Proof. Let us give some increments δm, δa to the sought-for variables m, a, i.e. consider the functions $m(t)+\delta m(t)$, $a(t)+\delta a(t)$, $t\in[t_0,T)$. We can use only the admissible variations $\delta m(t)$, $t\in[t_0, T)$, such that $m(t)+\delta m(t)$, $a(t)+\delta a(t)$ do not violate the conditions (6.52), (6.54) and (6.56). Using equation (6.52), it is easy to see that these increments are connected by the equality (6.83).

Let us determine the increment of functional (6.51) caused by the increment δm:

$$\delta I = I(m+\delta m) - I(m) = \int_{t_0}^{T} \rho(t)[\int_{a_{\max}(t)}^{t} l(\tau)\delta m(\tau)d\tau -$$

$$- \int_{a(t)}^{a(t)+\delta a(t)} l(\tau)[m(\tau)+\delta m_{\text{int}}(\tau)]d\tau] - P(t)\delta m(t)]dt =$$

$$= \int_{t_0}^{T} \rho(t)[\int_{a_{\max}(t)}^{t} [l(\tau) - l(a(t))]\delta m(\tau)d\tau + P(t)\delta m(t)]dt +$$

$$+ \int_{t_0}^{T} \rho(t) \int_{a(t)}^{a(t)+\delta a(t)} [l(a(t)) - l(\tau)][m(\tau)+\delta m(\tau)]d\tau dt$$

Finally, interchanging the order of integration in the first integral and introducing the notations $I'(t)$ as (6.81) and $\delta^2 I$ as (6.82) lead to formula (6.80). The lemma is thus proven.

Lemma 2. The convexity of *I*(*m*). If the function $l(t)$ is not increasing, then the abstract function $I(m)$ in OP (6.51)-(6.56) is convex on the set $\mathbf{U}_1$. If $l(t)$ is strictly decreasing, then the function $I(m)$ is strictly convex on the set $\mathbf{U}_1$.

The proof follows from the analysis of formula (6.82). By the statement of the mean we have

$$\delta^2 I = \int_{t_0}^{T} \rho(t)[l(a(t)) - l(a(t)+\chi(t))] \int_{a(t)}^{a(t)+\delta a(t)} [m(\tau)+\delta m(\tau)]d\tau dt$$

Let $m_1(\tau) > m_2(\tau)$, $\tau\in[t_0,T)$. Then $\delta m(\tau) = m_1(\tau) - m_2(\tau) > 0$, $\tau\in[t_0,T)$, and applying (6.83) we get $\delta a(t) \geq 0$, $t\in[t_0,T)$. Hence, $0 \leq \chi(t) \leq \delta a(t)$, $t\in[t_0,T)$, and $\delta^2 I \leq 0$. If $m_1(\tau) > m_2(\tau)$, $\tau\in[t_0,T)$, then $\delta a(t) \leq \chi(t) \leq 0$, $t\in[t_0,T)$, and $\delta^2 I \leq 0$.

Suppose now that $l(t)$ is strictly decreasing. Let the variation $\delta m(\tau) > 0$, $\tau\in\Delta_m\subset[t_0,T)$. Consider a set $\Delta := \{t\text{: } \Delta_m \cap [a(t), t] \neq \varnothing\}$. Then, in view of equality (6.83), the corresponding variation $\delta a(t)$, $t\in\Delta$, is such that

$m(\tau)+\delta m(\tau) \neq 0$, at least for some $\tau\in[a(t), a(t)+\delta a(t)]$. So, the integrand in the last formula for $\delta^2 I$ is negative for some values τ and $\delta^2 I\leq 0$. In so doing, $\delta a(t)$ can be finite for infinitesimal $\delta m(\tau)$ (then the function $I(m)$ is not differentiable but still convex).

The case $\delta m(.)<0$ is studied analogously. Lemma is thus proven.

Lemma 3. The necessary and sufficient conditions of extremum. In order for a function $m^*(t)$, $t\in[t_0, T)$, to be a solution of OP (6.51)-(6.56), it is necessary and sufficient that

$$\delta I = I(m^*+\delta m) - I(m^*+\delta m) \leq 0$$

for any admissible variation $\delta m(\tau)$, $\tau\in[t_0,T)$.

If $m^*(t)>0$, $t\in[t_0, T)$, then this necessary and sufficient extremum condition can be written in the form:

$$\begin{aligned} &I'(t) > 0 \quad \text{at} \quad m^*(t) = m_{\min}(t), \\ &I'(t) < 0 \quad \text{at} \quad m^*(t) = M(t), \\ &I'(t) \equiv 0 \quad \text{at} \quad m_{\min}(t) < m^*(t) < M(t), \quad t\in[t_0, T). \end{aligned} \tag{6.84}$$

Proof is based on standard optimal control theory reasoning (Rockafeller, 1970; Ioffe & Tichomirov, 1979; Vasiliev, 1981; Rubinov, 2000, etc.); see also Section 3.1.

Lemma 4. The monotonicity of $I(a)$. Let $l(t)$ be strictly decreasing, $a_1(t)=a_2(t)$ for $t\in[a_1(t_2), t_2]$, $a_1^{-1}(t_0)\leq t_2<T$.
If $a_1(t) < a_2(t)$ for $t\in[t_1, a_1(t_2)]$, $a_1(a_1(t_2))\leq t_1< a_1(t_2)$,
then $I'(a_1; t) < I'(a_2;t)$, $t\in [t_1, a_1(t_2)]$, and the converse.
If $a_1(t) < a_2(t)$ for $t\in(t_2, t_3]$, $t_2< t_3\leq a_1^{-1}(t_2)$,
then $I'(a_1; t) < I'(a_2;t)$, $t\in (t_2, t_3]$, and the converse.

Proof. Let us consider $t \in [t_1, a_1(t_2)]$. Then $a_1^{-1}(t) = a_2^{-1}(t)$ and

$$I'(a_2;t) = \int_t^{a_2^{-1}(t)} \rho(\tau)[l(t) - l(a_2(\tau))]d\tau + P(t)\rho(t) =$$

$$= \int_t^{a_1^{-1}(t)} \rho(\tau)[l(t) - l(a_2(\tau))]d\tau + P(t)\rho(t) =$$

$$= I'(a_1;t) + \int_t^{a_1^{-1}(t)} \rho(\tau)[l(a_1(\tau)) - l(a_2(\tau))]d\tau > I'(a_1;t)$$

for $a_1(t) < a_2(t)$, $t\in[t_1, a_1(t_2)]$.

Now consider the interval $(t_2, t_3]$. Let $a_1(t) > a_2(t)$ for $t\in(t_2, t_3]$, then $a_1^{-1}(\tau)<a_2^{-1}(\tau)$, $\tau\in[a_1(t_2), t_2]$. We obtain that

$$I'(a_2;t) = \int_t^{t_2} \rho(\tau)[l(t) - l(a_2(\tau))]d\tau \; +$$

$$+ \int_{t_2}^{a_2^{-1}(t)} \rho(\tau)[l(t) - l(a_2(\tau))]d\tau + P(t)\rho(t) =$$

$$= I'(a_1;t) + \int_t^{a_1^{-1}(t)} \rho(\tau)[l(a_1(\tau)) - l(a_2(\tau))]d\tau I \; +$$

$$+ \int_{a_1^{-1}(t)}^{a_2^{-1}(t)} \rho(\tau)[l(t) - l(a_2(\tau))]d\tau < I'(a_1;t)$$

The reverse is proven analogously. The lemma is thus proven.

Statement 1. The structure of the OP solution.

Let $T = \infty$, the function $l(t)$ strictly decrease and a function $a^{\sim}(t)$, $t\in[t_0,\infty)$, exist such that $I'(t) \equiv 0$ for $a(.) \equiv a^{\sim}(.)$ and the corresponding $m^{\sim}(.)$ satisfies (6.56). Then OP (6.51)-(6.56) has the unique solution (m^*, a^*) of the following form:

$$m^*(t) = \begin{cases} m_{\min}(t) \text{ or } M(t), & t \in [t_0, \mu), \\ m^{\sim}(t), & t \in [\mu, \infty), \end{cases}$$

$$a^*(t) = \begin{cases} a_{\text{bnd}}(t), & t \in [t_0, \mu), \\ a^{\sim}(t), & t \in [\mu, \infty), \end{cases}$$

where the function

$$a_{\lim}(t) = \begin{cases} a_{\min}(t) & \text{if} \quad a_0 > a^{\sim}(t_0), \\ a_{\max}(t) & \text{if} \quad a_0 < a^{\sim}(t_0), \end{cases}$$

is found from (6.52) at $m^*\equiv m_{\min}$ or $m^*\equiv M$, the function $m^{\sim}$ is found from (6.52) at $a\equiv a^{\sim}$, and the instant μ is determined from the condition $a_{\text{bnd}}(\mu)=a^{\sim}(\mu)$.

Proof. Since the constructed pair (m^*, a^*) satisfies all restrictions (6.52)-(6.56), it is admissible. Let us show that (m^*, a^*) satisfies Lemma 3.

Consider the case $a_0 < a^{\sim}(t_0)$. Then $m^*(t) = M(t)$, $a^*(t) = a_{\lim}(t) < a^{\sim}(t)$, $a_{\lim}^{-1}(t) < a^{\sim -1}(t)$ for $t \in [t_0, \mu)$;

$$I'(a^*;t) = \int_t^{\mu} \rho(\tau)[l(t) - l(a_{\lim}(\tau))]d\tau +$$

$$+ \int_{\mu}^{a_{\lim}^{-1}(t)} \rho(\tau)[l(t) - l(a^{\sim}(\tau))]d\tau + P(t)\rho(t) <$$

$$< \int_t^{a_{\lim}^{-1}(t)} \rho(\tau)[l(t) - l(a^{\sim}(\tau))]d\tau + P(t)\rho(t) <$$

$$< \int_t^{a^{\sim -1}(t)} \rho(\tau)[l(t) - l(a^{\sim}(\tau))]d\tau + P(t)\rho(t) = 0$$

for $t \in [t_0, \mu)$; $I'(a^*; t) \equiv 0$ for $t \in [\mu, \infty)$. Hence, the necessary and sufficient conditions of extremum hold, and the pair (m^*, a^*) is a solution of the OP. Its uniqueness follows from Lemma 4.

The case $a_0 > a^{\sim}(t_0)$ is investigated similarly. Statement 1 is proven.

Statement 4. Turnpike theorem in normal form. Let:

1. the given function $l(t)$ be strictly decreasing, $l'(t) < 0$ be non-decreasing;
2. the given function $M(t)$ be bounded at $t \to \infty$;
3. the set of the solutions $a_T(t)$, $t \in [t_0, T]$, of equation (6.58) approach a unique trajectory $a^{\sim}(t)$, $t \in [t_0, \infty)$, as $T \to \infty$, such that $da^{\sim}/dt > 0$ and $a^{\sim}(t) \to \infty$ at $t \to \infty$;
4. the function $m^{\sim}(t)$ defined by $a^{\sim}(t)$ from equation (6.52) satisfy (6.56);
5. $K'(t) > 0$, $K(t) \to \infty$ as $t \to \infty$ or $a^{\sim}(t_0) > a_0$, $m_0(\tau) \geq d_0 > 0$, $\tau \in [\tau_0, t_0]$.

Then, when $T \to \infty$, the solution $a^*(.)$ of OP (6.51)-(6.56) tends to $a^{\sim}(.)$ on an asymptotically largest part of the interval $[t_0, T]$, e.g. for any $\varepsilon > 0$ the time T_0 exists such that for any $T \geq T_0$ the condition $|a^*(t) - a^{\sim}(t)| < \varepsilon$ is true on some subset $\Delta \subset [t_0, T]$ such that $\text{mes}(\Delta)/(T - t_0) \to 1$ for $T \to \infty$.

The Proof. Let us construct the following pair of functions $(m^\wedge, a^\wedge)$:

$$a^\wedge(t) = \begin{cases} a_{bnd}(t), & t \in [t_0, \mu), \\ a^{\sim}(t), & t \in [\mu, \theta), \\ a_{\min}(t), & t \in [\theta, T], \end{cases} \tag{6.85}$$

$$m^\wedge(t) = \begin{cases} m_{\min}(t) \text{ or } M(t), & t, \in [t_0, \mu), \\ m^{\sim}(t), & t \in [\mu, \theta), \\ m_{\min}(t), & t \in [\theta, T]. \end{cases}$$

Here the functions $a_{bnd}(t)$, $a_{min}(t)$, $a_{max}(t)$, are defined in Statement 1 and the instant μ is uniquely determined from the condition $a^{\sim}(\mu) = a_{min}(\mu)$ for the case $a_0 < a^{\sim}(t_0)$ and from $a^{\sim}(\mu) = a_{max}(\mu)$ for $a_0 > a^{\sim}(t_0)$. The instant Θ is uniquely found from the following conditions:

(1) $I'(\Theta) = 0$ at $m^*(t) \equiv m_{min}(t)$, $t \in (\Theta, T]$;

(2) $a^{\sim}(\Theta) \equiv a_{min}(\Theta)$.

Under these conditions, the values $(T-\Theta)/T \to 0$ and $(T-a^{\sim}(\Theta))/T \to 0$, the instant $\mu < \Theta$ exists and does not depend on the value T when $T \to \infty$.

Since the pair $(m^\wedge, a^\wedge)$ satisfies all restrictions (6.52)-(6.56), it is an admissible solution. By the construction of $a^\wedge$, one can see that

$$I'(t) \equiv 0, \quad t \in [\mu, a^{\sim}(\Theta)). \tag{6.86}$$

Let us analyze the corresponding variation of the functional I. In view of Lemma 1 and (6.86) we have

$$\delta I = \int_{t_0}^{\mu} I'(t)\delta m(t)dt + \int_{a^{\sim}(\theta)}^{T} I'(t)\delta m(t)dt + \delta^2 I.$$

By applying Lemma 4 for $\mu = a^{\sim}(t_2)$, we obtain that $I'(t) < 0$ for $\delta m(t) < 0$, $t \in [t_0, \mu)$, and $I'(t) > 0$ for $\delta m(t) > 0$, $t \in [t_0, \mu) \cup [\Theta, T]$. So, $I'(t)\delta m(t) > 0$, $t \in [t_0, \mu) \cup [\Theta, T]$, for any admissible variations δm and

$$\delta I > \int_{a^{\sim}(\theta)}^{\theta} I'(t)\delta m(t)dt + \delta^2 I. \tag{6.87}$$

Note that $a^\wedge(t) < a^{\sim}(t)$, $t \in [\Theta, T]$, and $I'(a^{\sim}; t) = 0$, $t \in [a^{\sim}(\Theta), \Theta]$. Then, by applying Lemma 4 for $\mu = t_2$, we obtain $I'(a^\wedge; t) < 0$, $t \in [a^{\sim}(\Theta), \Theta]$. Next, using (6.83) we have

$$\begin{aligned} &\int_{a(\theta)}^{\theta} I'(t)\delta m(t)dt = I'(\varsigma)\int_{a(\theta)}^{\theta} \delta m(t)dt = \\ &= I'(\varsigma)\int_{a(\theta)}^{a(\theta)+\delta a(\theta)} [m(\tau) + \delta m_{\text{int}}(\tau)]d\tau > \\ &> I'(\varsigma)M(\theta)\delta a(\theta), \quad a(\theta) \le \varsigma \le \theta. \end{aligned} \tag{6.88}$$

Since $I'(\zeta)<0$, only the variations $\delta a(\Theta)\geq 0$ can lead to $\delta I<0$. On the other hand, in view of the first condition of the statement we find that

$$I'(\varsigma) > \int_{\varsigma}^{a^{\sim-1}(\varsigma)} \rho(\tau)[l(\varsigma) - l(a^{\sim}(\tau))]d\tau >$$

$$> \rho(\varsigma)\int_{a^{\sim}(\varsigma)}^{\varsigma} [l(\varsigma) - l(u))] / a^{\sim\prime}(u)du >$$

$$> \rho(\varsigma)[\varsigma - a^{\sim}(\varsigma)]^2 l'(a^{\sim}(\varsigma)) / a'_{\min} >$$

$$> \rho(a^{\sim}(\theta))[\varsigma - a^{\sim}(\varsigma)]^2 l'(a^{\sim}(a^{\sim}(\theta))) / a'_{\min} >$$

$$> \rho(a^{\sim}(\theta))l'(a^{\sim}(a^{\sim}(\theta)))R, \tag{6.89}$$

where the constants are $R := [\zeta - a^{\sim}(\zeta)]^2 / a^{\sim\prime}_{\min} > 0$,

$a^{\sim\prime}_{\min} := \min\{a^{\sim\prime}(u),\ u \in [a^{\sim}(\zeta), \zeta]\}$.

By the third condition of the statement the value $a^{\sim}(\Theta)/T \leq (\Theta - a^{\sim}(\Theta))/T \to 0$ and $(\zeta - a^{\sim}(\zeta))/T \to 0$ for $T\to\infty$. Hence, $R/T\to 0$ for $T\to\infty$.

Consider now any admissible pair (m^n, a^n) such that the trajectory $a^n(t)>0$, $t\in[t_0, T]$, does not tend to $a^{\sim}(.)$ in the sense indicated in this statement. Then for any $T>t_0$ a value ε and a subset Δ of the interval $[t_0, T]$ exist such that $|a^n(t) - a^{\sim}(t)|>\varepsilon$, $t\in\Delta$, and $\mathrm{mes}(\Delta)>\kappa(T-t_0)$ for $T\to\infty$, $\kappa=\mathrm{const}>0$. Since $(T - a^{\sim}(\Theta))/T\to 0$ for $T\to\infty$, hence, $\mathrm{mes}(\Delta^{\sim})\to \mathrm{mes}(\Delta)$ for $T\to\infty$, where $\Delta^{\sim} := \Delta\cap[\mu, a^{\sim}(\Theta)]$. So, we obtain that

$$\int_{\mu}^{a^{\sim}(\theta)} [\delta a(t)]^2 dt = \int_{\mu}^{a^{\sim}(\theta)} [a^n(t) - a^{\sim}(t)]^2 dt > \varepsilon^2 mes(\Delta) \xrightarrow[t\to\infty]{} \infty.$$

Consider the second variation $\delta^2 I$. In view of Lemma 1 and the first condition of the statement we have

$$\delta^2 I \geq \int_{t_0}^{T} \rho(t)l'(t)\int_{a^{\sim}(t)}^{a^{\sim}(t)+\delta a(t)} [a(t) - \tau][m(\tau) + \delta m(\tau)]d\tau dt >$$

$$> \rho(a^{\sim}(\theta))l'(a^{\sim}(a^{\sim}(\theta))) \times$$

$$\times \int_{\mu}^{a^{\sim}(\theta)} \int_{a^{\sim}(t)}^{a^{\sim}(t)+\delta a(t)} [a(t) - \tau][m(\tau) + \delta m(\tau)]d\tau dt. \tag{6.90}$$

First, assume that $m^\wedge(\tau)+\delta m(\tau)\geq d_m>0$ for $\tau\in[t_0, a^{\sim}(\Theta))$. Then

$$\delta^2 I \geq -d_m \rho(a^{\sim}(\theta))a^{\sim\prime}(a^{\sim}(a^{\sim}(\theta)))\int_{\mu}^{a^{\sim}(\theta)}[\delta a(t)]^2 dt/2.$$

Using (6.87)-(6.90), we obtain the following inequality:

$$\delta I > \rho(a^{\sim}(\Theta))\, l'(a^{\sim}(a^{\sim}(\Theta)))[-d_m\varepsilon^2 \text{ mes } \Delta^{\sim} + RM(\Theta)a^{\sim}(\Theta)],$$

where $l'(.)<0$, $a^{\sim}(\Theta)/T\to 0$, $R/T\to 0$, $\text{mes}(\Delta^{\sim})>\kappa(T-t_0)$ for $T\to\infty$.

Hence, $\delta I > 0$ for all $T>T^*$, starting from some large T^*. This implies that the inequality $I(m^n, a^n) > I(m^\wedge, a^\wedge)$ is valid and the pair (m^n, a^n) cannot be a solution of the optimization problem.

Now assume that $m^\wedge(\tau)+\delta m(\tau)=0$ for $\tau\in\Delta_1\subset[t_0, a^{\sim}(\Theta))$. It is impossible in the first case of the fifth condition of the statement (then $m_{\min}(t) = K'(t)>0$, $\tau\in[t_0, T)$) but it can be valid in the second case of this condition. If Δ_1 is of a finite length or $\text{mes}(\Delta_1)/T \to 0$ at $T\to\infty$, than

$$\delta^2 I \geq -d_m \rho(a^{\sim}(\theta))l'(a^{\sim}(a^{\sim}(\theta)))\int_{[\mu,a^{\sim}(\theta)]-\Delta_1}[\delta a(t)]^2 dt/2$$

and all the above analysis is applicable.

Let $\text{mes}(\Delta_1)\to\infty$ for $T\to\infty$ such that $\text{mes}(\Delta_1)>\kappa_1(T-t_0)$ for $T\to\infty$, $\kappa_1=\text{const}>0$. Then, by the fifth condition,

$m^\wedge(t)\geq d_m>0$ for $t\in[t_0, \mu]$,

$m^\wedge(t)=K'(t)+m(a^\wedge(t))a^{\wedge\prime}(t)\geq K'(t)+\min\{d_0, d_m\}a^{\wedge\prime}(t)>0$ for $t\in[\mu,\Theta]$.

Hence,

$$\int_{t_0}^{T} m^\wedge(t)dt \xrightarrow[T\to\infty]{} \infty,$$

$$\int_{t_0}^{a^{\sim}(\theta)} |\,\delta m(t)\,|\, dt > \int_{\Delta_1} |\,m^\wedge(t)\,|\, dt \xrightarrow[T\to\infty]{} \infty. \tag{6.91}$$

Then by the theorem of the mean and (6.83), we obtain from (6.90) that

$$\delta^2 I > -\int_{t_0}^{a^{\sim}(\theta)} \rho(t) l'(a^{\sim}(t))[a(t) - \varsigma(t)] \int_{a(t)}^{t} \delta m(\tau) d\tau dt >$$

$$> -\rho(a^{\sim}(\theta)) l'(a^{\sim}(a^{\sim}(\theta))) \int_{t_0}^{a^{\sim}(\theta)} \delta m(t) \int_{t}^{a^{-1}(t)} [a(\tau) - \varsigma(\tau)] d\tau dt$$

$$> -d_{\varsigma} \rho(a^{\sim}(\theta)) l'(a^{\sim}(a^{\sim}(\theta))) \int_{t_0}^{a^{\sim}(\theta)} \delta m(t) [a^{-1}(t) - t)] dt.$$

since $\delta m(t)>0$ for $t \in \Delta_1$ and, in view of (6.83), $\zeta(t) - a(t) \geq d_{\zeta}>0$.

Finally, using (6.87)-(6.91) we obtain the following inequality:

$$\delta I > \rho(a^{\sim}(\Theta))\, l'(a^{\sim}(a^{\sim}(\Theta)))[-d_{\zeta}Q + RM(\Theta)a^{\sim}(\Theta)],$$

where $l'(.)<0$, $a^{\sim}(\Theta)/T \to 0$, $R/T \to 0$, $Q/T \to \infty$ for $T \to \infty$. Hence, $\delta I > 0$ for all $T > T^*$, starting from some large T^*. This implies that the inequality $I(m^n, a^n) > I(m^\wedge, a^\wedge)$ is valid and the pair (m^n, a^n) cannot be a solution of the optimization problem in this case as well. Hence, any OP solution, if it exists, tends to $a^{\sim}(.)$ in the sense indicated in this statement.

Statement 4 is thus proven.

* * * * *

Thus, normal and strongest-form turnpike theorems have been established for the problem of equipment useful life optimization under technological change. They demonstrate that the optimal useful life of equipment comes close to a turnpike trajectory on a larger part of planning horizon. Such properties are well known for other optimization problems and play a significant role in economic theory.

The practical importance of the established results lies in finding some "efficient" trajectory with a simple structure (a turnpike) close to the optimal equipment useful life. In real situations, finding and investigating a turnpike trajectory is much easier than solving an original optimization problem.

The turnpike equipment useful life appears to be determined by the dynamics of technological change only and does not depend of the capital stock and initial equipment age structure. So, the turnpike regimes indicate some basic patterns of technological renovation. They can be used as a basis for decision-making in the rational replacement of industrial equipment.

Chapter 7

Appendix

1. MISCELLANEOUS FACTS OF ANALYSIS

This section contains several results of calculus and functional analysis that are repeatedly used in the monograph.

1.1 Vector and Integral Calculus

Let us consider the rectangular coordinate system x_1,x_2,x_3 in the three-dimensional space $\boldsymbol{R}^3$.

1.1.1 Gradient, Divergence and Rotation

The vector

$$\nabla S = \text{grad}\, S = \frac{\partial S}{\partial x_1}\mathbf{i} + \frac{\partial S}{\partial x_2}\mathbf{j} + \frac{\partial S}{\partial x_3}\mathbf{k}$$

is called the *gradient* of a scalar function $S(x_1,x_2,x_3)$. It defines the direction and magnitude of the maximum rate of increase of the function S at the point $\mathbf{x}=(x_1,x_2,x_3)$.

The scalar

$$\nabla \cdot \mathbf{V} = \text{div } \mathbf{V} = \frac{\partial V_1}{\partial x_1} + \frac{\partial V_2}{\partial x_2} + \frac{\partial V_3}{\partial x_3}$$

is called the *divergence* of a vector function $\mathbf{V}(x_1,x_2,x_3)=V_1\mathbf{i}+V_2\mathbf{j}+V_3\mathbf{k}$.

The scalar

$$\Delta S = \text{div grad } S = \nabla \cdot (\nabla S) = \nabla^2 S = \frac{\partial^2 S}{\partial {x_1}^2} + \frac{\partial^2 S}{\partial {x_2}^2} + \frac{\partial^2 S}{\partial {x_3}^2}$$

is called the *Laplacian* of a scalar function $S(x_1,x_2,x_3)$.

The vector

$$\nabla \times \mathbf{v} \equiv \nabla \wedge \mathbf{v} \equiv \text{rot } \mathbf{v} \equiv \text{curl } \mathbf{v} =$$

$$= \left(\frac{\partial v_3}{\partial x_2} - \frac{\partial v_2}{\partial x_3} \right)\mathbf{i} + \left(\frac{\partial v_1}{\partial x_3} - \frac{\partial v_3}{\partial x_1} \right)\mathbf{j} + \left(\frac{\partial v_2}{\partial x_1} - \frac{\partial v_1}{\partial x_2} \right)\mathbf{k}$$

is called the *rotation* (or *curl*) of a vector function $\mathbf{v}(x_1,x_2,x_3)$.

If $\nabla \times \mathbf{v} = 0$, then the motion (flow) is called *irrotational.*
The following properties are always true:

$$\nabla \cdot (\nabla \times \mathbf{v}) = 0, \qquad \nabla \times (\nabla S) = \mathbf{0}.$$

The gradient, divergence and rotation are independent of the choice of rectangular coordinates.

1.1.2 Gauss Divergence Theorem.

For a closed region with the volume R and the boundary surface $S=\partial R$:

$$\iiint_R \text{div}(\mathbf{V})dx_1dx_2dx_3 = \iint_{\partial R} \mathbf{V}ds = \iint_{\partial R} (\mathbf{V} \cdot \hat{\mathbf{n}})ds.$$

1.1.3 Dubois-Reymond's Lemma.

If $f(\mathbf{x})$ is continuous in $\mathbf{x}$ from a space domain $D \subset \boldsymbol{R}^3$ and

$$\iiint_R f(\mathbf{x})dx_1dx_2dx_3 = 0$$

for every region R contained in D, then $f(\mathbf{x}) \equiv 0$ for $\mathbf{x} \in D$.

1.1.4 Leibniz's Formula for Derivatives

If $f(x,t)$ and its partial derivative $\partial f(x,t)/\partial x$ are continuous in x,t; and functions $a(x)$ and $b(x)$ are differentiable,
then

$$\frac{d}{dx}\int_{a(x)}^{b(x)} f(x,t)dt =$$

$$= \int_{a(x)}^{b(x)} \frac{\partial f(x,t)}{\partial x}dt + f(x,b(x))\frac{db(x)}{dx} - f(x,a(x))\frac{da(x)}{dx}$$

1.2 Functional Spaces

Some well-known notations for the *functional spaces* used in this monograph are:

$\boldsymbol{C}_{[a,b]}$ - the space of all continuous functions defined on the interval $[a,b]$;

$\boldsymbol{C}^1{}_{[a,b]}$ - the space of all functions with a continuous derivative defined on the interval $[a,b]$;

$\boldsymbol{L}^\infty{}_{[a,b]}$ - the space of all functions defined and bounded almost everywhere on the interval $[a,b]$;

$\boldsymbol{BV}[a,b]$ - the space of all functions of bounded variation on the interval $[a,b]$.

The functional $\Phi(.)$ is an operator that puts a real value from $\mathbf{R}^1$ to each function $x(.)$ from a certain functional space $\boldsymbol{\Omega}$.

1.3 Calculus of Variations and Euler Equations

Calculus of variations is a classic mathematical technique that was developed over two hundred years mostly for geometrical, mechanical and physical applications. A *variational problem* consists of optimization of a certain functional on a set of smooth functions in a closed domain.

Further extension of the variational techniques to the case of non-smooth sought-for functions and domains leads to modern *optimal control theory* and its main tool - *principle of maximum* (Pontryagin et al, 1962).

There are many statements of variational problems (more or less general). Here we formulate the following classic problem in a two-dimensional space (Courant and Hilbert, 1953):

Problem. Find an extremum of the integral

$$\iint_G F(x_1, x_2, u(x_1, x_2), u_{x_1}(x_1, x_2), u_{x_2}(x_1, x_2))dx_1 dx_2 \rightarrow \text{extr} \qquad (7.1)$$

over the region $G \subset \boldsymbol{R}^2$ by determining a suitable function $u(x_1, x_2)$ which is continuous, has continuous first and second derivatives and takes on prescribed values on the boundary of G.

The *Euler equation* delivers a necessary condition of extremum for problem (7.1):

$$\frac{\partial F}{\partial u} - \frac{\partial}{\partial x_1}\left(\frac{\partial F}{\partial u_{x_1}}\right) - \frac{\partial}{\partial x_2}\left(\frac{\partial F}{\partial u_{x_2}}\right) = 0. \qquad (7.2)$$

Equation (7.2) is a partial differential equation of the second order with respect to the sought-for function $u(x_1, x_2)$.

In a one-dimensional case of Problem (7.1), the *Euler equation* is reduced to the following ordinary differential equation of the second order with respect to the sought-for function $u(x)$:

$$\frac{\partial F}{\partial u} - \frac{\partial}{\partial x}\left(\frac{dF}{du'}\right) = 0. \qquad (7.3)$$

2. MATHEMATICAL MODELS AND EQUATIONS

The following notes about various types of mathematical models and their relations are not exhaustive and cover only modelling tools used in the monograph.

2.1 Classification of Mathematical Models

We have used stochastic and deterministic models, continuous and discrete models, linear and nonlinear models, difference, differential and integral equations.

2.1.1 Deterministic and Stochastic Models

Stochastic (statistical) models involve connections among stochastic (probabilistic) characteristics of systems and processes under study. They are very useful for the analysis of repetitive processes but require a large amount of data to start modelling. A comprehensive analysis of all available information should be the first step of the system analysis.

Deterministic models include all models that operate with some quantitative characteristics without assuming their probabilistic nature. Deterministic models are helpful in many realistic situations where there are relatively few sources of uncertainty inside the system. If stochastic fluctuations of the characteristics are absent or can be neglected, then the system can be described in terms of the deterministic models.

In many cases, deterministic models deal with averaged probabilistic characteristics of processes under study (an average "concentration of pollutant" instead of the real concentration, the expected value of "equipment lifetime" instead of the real "equipment lifetime", and so on) and are based on the approximation of a real process.

Some engineering systems belong to complex systems with high dimensionality and uncertainty of relationships inherent in them. Nevertheless, deterministic models are commonly used for their description. The reason lies in the increasing complexity of mathematical description using stochastic factors without substantial insight in the interpretation of a process. Implementation of some processes is unique and accompanied by a shortage of data (especially, for large-size systems). It is important to remember that in modelling of integrated systems the roughest model block determines the total error of modelling (and the corresponding quality of prediction and control).

The majority of models in this monograph are deterministic. Stochastic models are used in Chapter 5 because the diffusion processes modeled there have a substantially stochastic nature.

2.1.2 Continuous and Discrete Models

Depending on a type of used data and techniques of process description, the mathematical models are subdivided into *continuous* and *discrete models*. *Continuous models* operate with continuous variables, and *discrete models* operate with discrete variables.

A discrete model uses vectors like $\mathbf{x} = (x_1, x_2, ..., x_n) \in \mathbf{R}^n$ as dependent (sought-for) and independent variables. A general form of discrete models is

$$F_j(x_1, x_2, ..., x_n) = 0, \qquad j=1,...,m, \tag{7.4}$$

where $F_j(.)$ are some functions of n scalar variables.

A continuous model operates with scalar- or vector-valued functions $y(x)$ of an independent continuous variable x (scalar or vector) defined on some domain $D\subset\mathbf{R}^n$, $n\geq1$. Continuous dynamic models include time t as one of the independent variables. A general form of continuous models is

$$\Phi(y) = 0, \tag{7.5}$$

where $\Phi(.)$ is a functional that sets a real value for each function $y(.)$ from a certain functional space Ω (see Section 1.2 above).

A discrete analogue can usually be constructed for a continuous model, and vice versa. For the most of continuous models considered above, their discrete analogues are known and commonly used in numeric algorithms of model investigation. The choice between continuous and discrete models depends on model's capabilities in reflecting the problem and process under investigation.

2.1.3 Linear and Nonlinear Models

The choice between linear and nonlinear models depends on nature of the process under study or on the desired level of process approximation.

Many real-life processes are nonlinear but are commonly described by approximate linear models because the linear models are simpler and have a better theory or investigation technique. Other processes are substantially nonlinear and their linearization leads to oversimplified description.

Linear discrete model in a general form is the well-known *system of linear algebraic equations* (*SLAE*) :

$$\sum_{i=1}^{n} a_{ij} x_j = b_i, \quad i = 1, \ldots, m,$$

or

$$A\mathbf{x} = \mathbf{b}, \tag{7.6}$$

where $\mathbf{x}=(x_1, x_2, \ldots, x_n) \in \mathbf{R}^n$, $\mathbf{b}=(x_1, x_2, \ldots, x_m) \in \mathbf{R}^m$, and $A = \{a_{ij}\}$ is an $m \times n$ *matrix*.

Model (7.6) represents a convenient and completely investigated mathematical object. If $m=n$ and the determinant $\det A \neq 0$, then SLAE (7.6) has a unique solution $\mathbf{x}$ (under given A and $\mathbf{b}$) that can be found by various fast algorithms.

Linear continuous model is the model (7.5) with a linear functional $\Phi(.)$. The linear functional preserves the *linear operations of addition and scalar multiplication* for all elements x, y of the functional space Ω:

$$\Phi(y+z) = \Phi(y) + \Phi(z), \qquad \Phi(\alpha y) = \alpha \Phi(y) \text{ for } \alpha \in \mathbf{R}^1.$$

Nonlinear continuous model is the equality (7.5) when the functional $\Phi(.)$ is nonlinear.

Nonlinear discrete models of the form (7.4) with nonlinear functions $F_j(.)$ do not possess a general theory, and investigation of a specific system of nonlinear equations often runs into great theoretical or numeric difficulties. The solution may be non-unique or not exist in general. However, there are special classes of nonlinear discrete models, which have well-developed theory and applications. One of them - difference equations - is used in the book and is mentioned below.

2.1.4 Difference, Differential and Integral Models

Difference Equations represent a special class of the discrete model (7.4) and correspond to differential equations in continuous models (see below). They naturally arise in the development of numeric methods for differential equations as well as in direct description of some applied dynamic systems. The major specific of difference equations is that they connect the values of sought-for variables x_1, x_2, ..., x_n in several neighboring points only. The difference equations may be linear or *nonlinear*.

Applied to dynamic processes, the difference models describe processes in a discrete time when the future dynamics depends on one or several previous states of the process only. Such simple nonlinear difference models might have a very complicated dynamical behaviour; some cases are exposed in Chapter 5 applied to population dynamics.

Depending on type of the functional $\Phi(.)$ in the general continuous model (7.5), there are two major types of continuous models: *differential* and *integral models.*

Differential Equations represent functional connections between sought-for functions and some of their derivatives. A general form of such equations is:

$$F(x_1,\ldots,x_n,\frac{\partial y}{\partial x_1},\frac{\partial y}{\partial x_2},\ldots,\frac{\partial y}{\partial x_n},\frac{\partial^2 y}{\partial x_1{}^2},\frac{\partial^2 y}{\partial x_1 \partial x_2},\ldots)=0. \qquad (7.7)$$

where $F(.)$ is a nonlinear function of several variables.

In general case, expression (7.7) is a *partial differential equation* (*PDE*). If the sought-for function $y(x)$ depends only on one independent variable t, then (7.7) is an *ordinary differential equation* (*ODE*).

The differential equation (7.7) is *linear* if the function $F(.)$ is linear with respect to the sought-for functions and their derivatives.

If the functions F and y are vector-valued, $F(.)=\{F_1(.),\ldots,F_n(.)\}$, $y(x)=\{y_1(x),\ldots,y_n(x)\}$, then the expression (7.7) represents a system of n differential equations with respect to $y_1,\ldots,y_n$.

Applied to *dynamical processes* (i.e., processes developed in time t), differential models describe a special class of such processes (*processes without after-effect*) when the dynamics of future development depends on a current state of the process only. Such approximation appears to be good enough for many physical, mechanical, and other processes. The reason is that various initial perturbations quickly vanish in real processes and may be excluded from a process model.

Classification of Partial Differential Equations. Linear partial differential equations of the second order represent the most frequently used spatial models in engineering applications. A general form of the linear PDE of the second order in a two-dimensional space with respect to the unknown $u(x,y)$ is

$$A\frac{\partial^2 u}{\partial x^2}+2B\frac{\partial^2 u}{\partial x \partial y}+C\frac{\partial^2 u}{\partial y^2}+D\frac{\partial u}{\partial x}+E\frac{\partial u}{\partial y}+F\,u+G=0$$

A well-known classification of the linear PDEs of the second order in terms of the above general equation is given in Table 7.1:

Table 7.1. Classification of the linear partial differential equations of the second order.

Condition	Type of equation	Example
$B^2 < AC$	*Elliptic*	*Laplace equation* *Poisson equation*
$B^2 = AC$	*Parabolic*	*Diffusion equation* *Schrödinger equation*
$B^2 > AC$	*Hyperbolic*	*Wave equation*

Another useful general presentation of *nonstationary* linear PDEs of the second order in two space variables x, y and time t

$$\frac{\partial^2 u}{\partial x^2}+\frac{\partial^2 u}{\partial y^2}=a+b\,u+c\,\frac{\partial u}{\partial t}+d\frac{\partial^2 u}{\partial t^2}$$

includes the following well-known special equations used in the current book:

Table 7.2. Major linear partial differential equations of the second order.

Case	Name of the equation	The equation
a=0; b=0; c=0; d=0	Laplace equation	$\frac{\partial^2 u}{\partial x^2}+\frac{\partial^2 u}{\partial y^2}=0$
a≠0; b=0; c=0; d=0	Poisson equation	$\frac{\partial^2 u}{\partial x^2}+\frac{\partial^2 u}{\partial y^2}=-4\pi\rho$
a=0; b≠0; c=0; d=0	Helmholtz equation	$\frac{\partial^2 u}{\partial x^2}+\frac{\partial^2 u}{\partial y^2}=-k^2 u$
a=0; b=0; c≠0; d=0	Diffusion equation	$\frac{\partial^2 u}{\partial x^2}+\frac{\partial^2 u}{\partial y^2}=c\,\frac{\partial u}{\partial t}$
a=0; b=0; c=0; d≠0	Wave equation	$\frac{\partial^2 u}{\partial x^2}+\frac{\partial^2 u}{\partial y^2}=\frac{1}{v^2}\frac{\partial^2 u}{\partial t^2}$

Integral Equations contain some integrals with sought-for functions in their integrands. In linear case, they provide the most general representation of deterministic continuous models. It is known that any linear functional $\Phi(x)$ of Model (7.5) can be presented in an integral form.

Let us consider the case of a scalar one-dimensional independent variable t. Then, in general situation, a linear functional $\Phi(x)$ with respect to the function $x(t)$, $t\in[t_0, T]$, can be described by the following integral

$$\Phi(x)=\int_a^b K(\tau)x(\tau)d\tau,$$

where $K(t)$ is a given function. Then the continuous model (7.5) of a linear system can be described as the integral model:

$$\int_a^b K(t,\tau)x(\tau)d\tau = f(t), \quad t \in [a,b]. \tag{7.8}$$

where $f(t)$ is a given function.

Formula (7.8) describes the *Fredholm integral equation of the first kind* with respect to the sought-for $x(t)$, $t \in [t_0, T]$.

The *Fredholm integral equation of the second kind*

$$x(t) = \int_a^b K(t,\tau)x(\tau)d\tau + f(t), \quad t \in [a,b], \tag{7.9}$$

appears to be more common as compared with the Fredholm integral equation of the first kind (7.8). It represents a direct generalization of the discrete linear model in the form of SLAE (7.6).

After the discretization on the variable t, the equation (7.9) leads to the SLAE (7.6). The analogy between continuous integral models and their discrete analogues (SLAE) is very useful for better understanding and interpretation of a model. However, theory of linear continuous models is more complex as compared with linear discrete models. In particular, a significant difference exists between integral equations of the first and the second kinds.

Now let us consider dynamical (casual) systems. The current state of a dynamical system (process) depends on the past states only (it cannot depend on the future). Hence, $K(t,\tau)\equiv 0$ at $\tau > t$ in the model (7.9) for dynamical systems, and we obtain the following dynamical model:

$$x(t) = \int_a^t K(t,\tau)x(\tau)d\tau + f(t), \quad t \in [a,b], \tag{7.10}$$

which is known as the *Volterra integral equation of the second kind* with respect to the sought-for x.

The *Volterra integral equation of the first kind* is obtained similarly from the expression (7.8) and does not contain the sought-for $x(t)$ outside of the integral:

$$\int_a^t K(t,\tau)x(\tau)d\tau = f(t), \quad t \in [a,b].$$

Despite exterior similarity of the Fredholm and Volterra integral equations, their properties are quite different.

Nonlinear integral equations are generalizations of linear integral equations (7.8)-(7.10) with the integrand $K(\tau,t)x(\tau)$ replaced with a nonlinear function $F(x(\tau),\tau,t)$ of $x(\tau)$. Nonlinear Volterra integral equations are considered in more details in Section 2.2.

As opposed to differential models, the *dynamical integral models* take into account the *after-effect (persistence, contagion, hereditary effects)* when a continuous sequence of the past states of a dynamical system impacts the future evolution of the system.

Integral models are more general but differential models are simpler and more efficient during analytical and numerical study, so they are more common. A general *selection rule* is: if a differential model can efficiently describe a process with a required accuracy, then there is no need to construct and use an integral model.

There are even more general dynamical models, which are not considered in this book. For example, *functional equations with causal operators* (Corduneanu, 2002) encompass most types of deterministic dynamical models used in applied science and engineering: ordinary differential equations, Volterra integral equations, equations with delayed argument and Volterra integro-differential equations, and so on.

Right choice of the type of applied model is an important but difficult question. Table 7.3 provides a *classification and relationship matrix* that matches properties of different types of mathematical models with the features of real-life systems. Many of these model properties and system features have used and investigated somewhere in this monograph but some of them have not been mentioned and an interested reader is recommended to provide an additional search in scientific literature. This classification reflects the research experience of the authors.

Such classification cannot be exhaustive and exact enough. It is approximate and not complete and considers only *deterministic linear models* with an emphasis on dynamic systems. An important piece of the classification – algebraic equations – would be lost if we consider nonlinear models. The authors hope that Table 7.3 will be helpful and give an initial idea for the future model choice during investigation of applied systems and processes.

Table 7.3. Relations among real-life systems and various types of models.

Systems		**Models**			
		Continuous		**Discrete**	
Feature of system (process)	**Value of the feature**	**Integral equations**	**Differential equations**	**Algebraic equations (Ax=b)**	**Difference equations**
Causality	Dynamic (casual)	Volterra type (with variable upper limit)	Initial value problem	With a triangle matrix A	Initial value problem
	Non-dynamic (non-casual)	Fredholm type (with fixed limits)	Boundary value problem	With a full non-zero matrix A	Boundary value problem
Time-invariance	Autonomous	With convolution kernels	With constant coefficients	matrix A is constant	With constant coefficients
	Non-autonomous	With non-convolution kernels	With variable coefficients	matrix A depends time	With variable coefficients
Level of disorder and complexity	Regular, laminar	Of the first or the second kind	Regular	$\det A \neq 0$	Regular
	Singular, turbulent	Of the third kind	With singularities	$\det A = 0$ at some points	With quasi-chaotic regimes
Level of robustness and correctness	Well-defined	Of the second kind	Well-posed	Well-posed	Well-posed
	Ill-defined	Of the first kind	Ill-posed (stiff equations)	| maximal eigenvalue of A | >> | minimal eigenvalue of A |	Ill-posed (stiff equations)
Number of sought-for variables	=1	Scalar equation	One equation	One equation	One equation
	>1	Vector equation	System of equations	System of equations	System of equations
Dimension (number of independent variables)	=1	One-dimensional equations	Ordinary differential equations	Equations with matrices and vectors	Ordinary difference equations
	>1	Multi-dimensional equations	Partial differential equations	Equations with tensors	Equations with partial differences

2.2 Integral Dynamical Models and Volterra Integral Equations

In Chapters 4 – 6, we have extensively used integral dynamical models of the form

$$x(t) = \int_{a(t)}^{t} K(t,\tau)Y(\tau)x(\tau)d\tau, \tag{7.11}$$

Such models are naturally reduced to standard Volterra integral equations (VIE)

$$x(t) = \int_{t_0}^{t} K(t,\tau)Y(\tau)x(\tau)d\tau + f(t), \tag{7.12}$$

where $f(t) = \int_{a(t)}^{t_0} K(t,\tau)Y_0(\tau)x_0(\tau)d\tau$ is a given function on some time interval $[t_0, t_1]$ such that $a(t) \le t_0$, $t \in [t_0, t_1]$. Expression (7.12) represents an equation of the second kind with respect to $x(t)$ or of the first kind with respect to $Y(t)$.

Below we provide several known results about Volterra integral equations used in the monograph. We consider the following nonlinear Volterra integral equation:

$$x(t) = \int_{a}^{t} F(t,s,x(s))ds + f(t), \qquad t \in (a,b). \tag{7.13}$$

The special case $F(t,s,x) = K(t,s)G(x)$ of equation (7.13) is known as *Hammerstein-Volterra integral equation*:

$$x(t) = \int_{a}^{t} F(t,s,x(s))ds + f(t), \qquad t \in (a,b).$$

2.2.1 Solvability of Volterra Integral Equations

There are many results about existence and uniqueness of VIE solutions Volterra integral equations (Caljuk, 1977; Barton, 1983; Jerri, 1985; Gripenberg et al, 1990; Corduneanu, 1991). They differ in smoothness requirements for the given functions and the form of the equation. One of simple and natural criteria is:

Statement 1 (**Solution Existence**). If $f \in \boldsymbol{C}_{[a,b]}$, $F \in \boldsymbol{C}_{\Delta}$, where $\Delta = \Delta_a \times \{x: |x - f(t)| \leq r\}$, $\Delta_a = \{(t,s): a \leq s \leq t \leq b\}$, then there exists at least one solution $x \in \boldsymbol{C}_{[a,a+\delta]}$ of equation (7.13) where

$$\delta = \min\{\, b\text{-}a,\ r/M\},\quad M = \sup|\, F(t,\ s,\ x)| \text{ on } \Delta.$$

In general, $F(s,t,x)$ does not need to be continuous in t and s. We illustrate that fact in the following statement concerning a vector VIE (7.13) where the values of functions $x(.)$ and $F(.)$ are n-vectors: $F(t,s,x) \in R^n$, $x(t) \in R^n$.

Statement 2 (**Solution Existence**). If:

- $f \in \boldsymbol{C}_{[a,b]}$,
- F is measurable in t and s for $(t,s) \in \Delta_a$, $\Delta_a = \{(t,s): a \leq s \leq t \leq b\}$, and is continuous in x for each fixed (t,s),
- for each bounded $B \subset R^n$, there exists a nonnegative measurable function $m(t,s)$ such that

$$|F(t,s,x)| \leq m(t,s),\ \ (t,s) \in \Delta_a,\ x \in B, \text{ and}$$

$$\sup \int_a^t m(t,s)ds < \infty,\quad t \in [a,b],$$

- for any $\tau \in [a,b]$ and $x \in \boldsymbol{C}_{[a,b]}$, $|x(t)| \leq M$, the following limit is true

$$\sup \int_a^b |\, F(t,s,x(s)) - F(\tau,s,x(s))\,|\, ds \to 0 \quad \text{as} \quad t \to \tau,$$

then there is at least one solution $x \in \boldsymbol{C}_{[a,a+\delta]}$ of equation (7.13) for sufficiently small values $\delta > 0$.

After determining a continuous solution $x(t)$ of VIE (7.13) on a small interval $[a,a+\delta] \subset [a,b]$, $\delta < b\text{-}a$, we usually want to continue it on a larger interval. Then two distinct situations can occur:

- the solution $x(t)$, $t \in [a,\ a+\delta]$, is *continuable* (or *unsaturated*) and can be extended to a larger interval $[a,\ a+\delta_1)$, $\delta < \delta_1$;

- the solution $x(t)$, $t\in[a,\ a+\delta)$, is *maximally defined* (or *saturated*) and cannot be extended to a larger interval.

In the latter case, the solution $x(t)$ is unbounded in a neighborhood of the point $a+\delta$, i.e., $\limsup |x(t)| = \infty$ as $t \to a+\delta$.

Solution Uniqueness. The key condition for the uniqueness of the solution x of VIE (7.13) is the Lipschitz condition for the function $F(t,s,x)$ with respect to x:

$$|F(t,s,x) - F(t,s,y)| \le L(t,s)|x-y|,\ \ (t,s)\in\Delta_a,\ x\in B,$$

Adding this condition to the previous statements guarantees that the VIE (7.13) has a unique solution x.

In the case of *linear VIE* $F(s,t,x) = K(s,t)x$, a unique solution $x(t)$ exists on the whole interval $[a,b]$ and may be defined by the following formula

$$x(t) = \int_a^t R(t,s)f(s)ds + f(t), \tag{7.14}$$

where the so-called *resolvent kernel* $R(t,s)$ is a solution of the following linear VIE:

$$R(t,s) = \sum\nolimits_{i=1}^{\infty} K_i(t,s), \tag{7.15}$$

$$K_1(t,s) = K(t,s), \quad K_i(t,s) = \int_s^t K_{i-1}(t,u)K(u,s)du, \quad i>1.$$

The resolvent kernel R itself is a solution of the following linear VIE:

$$R(t,s) = \int_s^t R(t,u)K(u,s)du + K(t,s).$$

2.2.2 Correctness and Stability of Volterra Integral Equations

The VIEs of the second kind under natural assumptions belong to *well-defined* (*well-posed*) problems in the sense that their solutions are stable with respect to small perturbations of the given functions. This property is also known as *continuous dependence of VIE solutions on the given data* and as *correctness by Hadamard* (in computational mathematics).

Let us define a perturbed VIE (7.13) as

$$y(t) = \int_a^t [F(t,s,y(s)) + \delta F(t,s,y(s))]ds + f(t) + \delta f(t), \qquad (7.16)$$

where the functions $\delta F(t,s,x)$, $\delta f(t)$ are small perturbations of the given functions $F(t,s,x)$, $f(t)$, and estimate the corresponding perturbation of the VIE solution. We have used the perturbation technique for an integral dynamical model in Chapter 5.

Statement 3 (**Solution Correctness**). Let $f \in \boldsymbol{C}_{[a,b]}$, $F(t,s,x)$ be continuous in t and s and locally Lipschitz in x in $\Delta = \Delta_a \times R$, $\Delta_a = \{(t,s)\colon a \le s \le t \le b\}$, $x(t)$ be the unique saturated solution of VIE (7.13) on $[a,b)$.

Then for every $\varepsilon > 0$ and $c < b$ the number $\delta = \delta(\varepsilon, c) > 0$ exists such that, if $\delta f \in \boldsymbol{C}_{[a,b]}$, $\delta F(t,s,x) \in \boldsymbol{C}_\Delta$ and

$$|\delta f(t)| \le \delta, \quad |\delta F(t,s,x)| \le \delta, \quad (t,s,x) \in \Delta,$$

then

$$|x(t) - y(t)| \le \varepsilon, \quad t \in [a,c],$$

for any solution $y(t)$, $t \in [a,c]$, of the perturbed VIE (7.16).

Thus, conditions of the correctness for the VIE of the second kind are essentially the same that of existence and uniqueness of VIE solutions. It is not true for VIE of the first kind. The incorrectness of the VIE of the first kind is similar to the problem of numeric differentiation. Thus, solutions of the linear VIE of the first kind

$$\int_a^t K(t,\tau)x(\tau)d\tau = f(t), \quad t \in [a,b]. \qquad (7.17)$$

are incorrect with respect to small *continuous* perturbations $\delta K(t,s)$, $\delta f(t)$ of the given functions $K(t,s)$, $f(t)$. However, they are correct with respect to small *continuously differentiable* perturbations δK, δf in the case of $|K(t,t)| \ge k > 0$, $t \in [a,b]$.

We would like to note that the Fredholm integral equations of the first kind possess another more essential type of incorrectness.

2.2.3 Stability of Volterra Integral Equations

The notion of stability of VIE solutions is similar to the correctness considered above but considers a VIE solution on an infinite interval. The definitions of stability for VIE solutions vary in different investigations. Let us illustrate this notion in a simple case.

Following (Vinokurov, 1969), the solution $x(t)$, $t\in[a,\infty)$ of VIE (7.13) is *stable with respect to the given function* $f(t)$ if for every $\varepsilon>0$ there is $\delta=\delta(\varepsilon)>0$ such that, if $\delta f\in \mathbf{C}_{[a,\infty)}$ and $|\delta f(t)|\le\delta$, $t\in[a,\infty)$, then

$$|x(t)-y(t)|\le\varepsilon,\quad t\in[a,\infty),$$

for any solution $y(t)$, $t\in[a,\infty)$, of the perturbed VIE (7.16) at $\delta F\equiv 0$. A sufficient stability condition for linear VIE is given below.

Statement 4 (**Solution Stability**). If $F(s,t,x) = K(s,t)x$, the *linear* VIE (7.13) has a unique solution $x(t)$ on the interval $[a,\infty)$, and the resolvent kernel $R(t,s)$ (7.15) satisfies the following condition:

$$\int_a^t \| R(t,\tau) \| d\tau \le \text{const} < \infty,\quad t\in[a,\infty),$$

then the solution x is stable with respect to the given function f.

Analogously, the stability of VIEs with respect to other given functions can be introduced and investigated. In the case of integral dynamical models of the form (7.11), it is natural to consider stability of the solution $x(t)$, $t\in[t_0,\infty)$, with respect to the perturbations of the given $x(t)=x_0(t)$ on the prehistory interval $[\min\{a(t),t_0\},\ t_0]$ (Yatsenko, 1991).

As opposed to the above well-posedness property, stability is not automatic and does not hold for many applied evolving systems.

2.2.4 Integral Inequalities

Integral inequalities play important role in the theory and numeric methods of differential and integral equations. A common example of general results is (Corduneanu, 1991):

Statement 5. (**Volterra integral inequality**) If VIE (7.13) has a unique continuous solution $x(t)$ on the interval $[a, b]$, the function $F(s,t,x)$ is monotonically nondecreasing in x, and a continuous function $z(t)$ satisfies the following inequality:

$$z(t) \leq \int_a^t F(t,s,z(s))ds + f(t), \qquad t \in [a,b], \tag{7.18}$$

then $z(t) \leq x(t)$, $t \in [a, b]$.

When we can evaluate a solution of the equation (7.13), many specific estimates in an explicit form follow from this statement. For example, if $F(s,t,x) = K(s,t)G(x)$ (Volterra-Hammerstein equation), then

$$z(t) \leq \Omega^{-1}\left\{\Omega[\sup_{a \leq \tau \leq t} f(\tau)] + \int_a^t \sup_{s \leq \tau \leq t} K(\tau,s)ds\right\},$$

$$\text{where } \Omega(u) = \int_0^u \frac{d\tau}{G(\tau)}$$

In the linear case $F(s,t,x) = U(s)x$, $f(t) \equiv f = \text{const}$, Statement 5 leads to the following well-known estimate often used in ODE theory:

Statement 6 (**Gronwall-Bellman Inequality**). If the function $U(s,t,x)$ is continuous and nonnegative, and

$$z(t) \leq C + \int_a^t U(s)z(s)ds, \qquad t \in [a,b],$$

then

$$z(t) \leq C\exp\left[\int_a^t U(s)ds\right], \qquad t \in [a,b].$$

In particular, the inequality $|x(t)| \leq e^{\|K\|(t-a)}\|f\|$ is valid for the linear VIE of the second kind (7.10).

References

Achenbach, J. *Wave Propagation in Elastic Solids,* Elsevier Science Publishers, Amsterdam, 1984.

Adler F.R. *Modeling the Dynamics of Life: Calculus and Probability for the Life Scientists*, Brooks/Cole Publishing Company, Pacific Grove/ Bonn/ Boston, 1998.

Agrafiotis, G.K. and Tsoukalas, M. Excess-Time Renewal Theory with Applications, *J. Oper. Res. Soc.*, **41**(1990), No.1, 69--82.

Aiba, S., Hamphrey A., Millis, N. *Biochemical Engineering,* Academic Press, New York, London, 1973.

Aistrakhanov, D., Ishchuk, V., and Yatsenko, Yu. The Turnpike Approach to Modeling Renewal Rates for Engineering Systems, *Engineering Simulation,* Gordon and Breach Sci. Publishers, **10**(1993), No.6, 941-954.

Akcasu, L., Lellouche, G. and Shotkin, L. *Mathematical Methods in Nuclear Reactor Dynamics*, Academic Press, New York, 1971.

Alekseev, V.M. Tikhomirov, V.M., and Fomin, S.V. *Optimal Control (Contemporary Soviet Mathematics)*, Plenum Publish. Corp., New York, 1987

Angell, T.S. and Olmstead, W.E. Singularly Perturbed Volterra Integral Equations, *SIAM J. Appl. Math.*, **47**(1987), 1-14.

Angell, T.S., On The Optimal Control Of Systems Governed By Nonlinear Volterra Equations, *J. Opt. Theory Appl.*, **19**(1976), 29-45.

Anger, B. and Portenier, C. *Radon Integrals,* Birkhäuser, Boston, 1992.

Anthony, M. and Biggs, N. *Mathematics for Economics and Finance: Methods and Modelling*, Cambridge University Press, 1996.

Antonelli, P., Ingarden R. and Matsumoto M. *The Theory of Sprays and Finsler Spaces with Applications in Physics and Biology*, Kluwer Academic Publishers, Dordrecht, 1993.

Arutjunian, N.H. and Kolmanovskii, V.B. *Creep Theory Of Nonhomogeneous Bodies*, Nauka, Moscow, 1983 (in Russian).

Azbelev, N.V. and Maksimov, V.P. Equations With Delayed Arguments, *Differential Equations,* **18**(1983), 1419-1441.

Bailey, J. E. Toward A Science Of Metabolic Engineering, *Science*, **252**(1991), 1668-1675.

Baker, C.T.H. *The Numerical Treatment of Integral Equations*, Clarendon Press, Oxford, 1977.

Bakke, V.L., A Maximum Principle For An Optimal Control Problem With Integral Constraints, *J. Opt. Theory Appl.*, **13**(1974), 32-55.

Banks, H.T., Kareiva, P.M., and Zia, L. Analyzing Field Studies of Insect Dispersal Using Two-Dimensional Transport Equations, *Environmental Entomology*, **17**(1988), 815-820.

Barnett, S. *Introduction to Mathematical Control Theory*, Oxford University Press, New York, 1975.

Bellomo, N. and Preziosi, L. *Modelling Mathematical Methods and Scientific Computation*, CRC Press, Boca Raton, 1994.

Belotserkovskii, S.M., Kochetkov A.A., Krasovskii, A.A., and Novitskii, V.V. *Introduction To Aeroautoelasticity*, Nauka, Moscow, 1980 (in Russian).

Bergh, J.C.J.M. van den, and Nijkamp, P. Dynamic Macro Modelling for Sustainable Development: Economic-Environmental Integration and Materials Balance, *Revue europeenne des sciences sociales*, **XXXI** (1993), No.96, 241-270.

Bergh, J.C.J.M. van den, *Ecological Economics and Sustainable Development: Theory, Methods and Applications*, Edward Elgar, Cheltenham, England, 1996.

Block, H.W., Borges, W.S., and Savits, T.H. A General Age Replacement Model with Minimal Repair, *Nav. Res. Logist.*, **35**(1988), 365-372.

Bloom, F. *Ill-Posed Problems For Integrodifferential Equations in Mechanics And Electromagnetic Theory*, SIAM, Philadelphia, 1981.

Bogoliubov, N. N. and Mitropolsky, Y. A. *Asymptotic Methods in the Theory of Nonlinear Oscillations*, Gordon and Breach, New York, 1961.

Boltzmann, L. Zur Theorie Der Elastischen Nachwirkung, *Sitz. Acad. Wiss. Wien.*, **70**(1874), 275-306.

Brekhovskikh, L.M. and Goncharov, V. *Mechanics of Continua and Wave Dynamics*, Springer-Verlag, New York, 1985.

Brems, H. *Quantitative Economics Theory: A Synthetic Approach*, Wiley, New York, 1967.

Brogan, W.L. *Modern Control Theory*, Quantum, New York, 1974.

Brown, M. *On the Theory and Measurement of Technological Change*, University Press, Cambridge, 1967.

Bruckner, E., Ebeling, W., Jimenez Montano, M.A. et al., *Technological Innovations - A Self-Organisation Approach*, WSB-Paper FS II 93-302, Berlin Science Center for Social Science Research, Berlin, 1993.

Brunner, H. and Houwen, P.J. van der, *The Numerical Solution of Volterra Equation*, North-Holland, Amsterdam, 1986.

Brunner, H. and Yatsenko, Yu. Spline Collocation Methods for Nonlinear Volterra Integral Equations with Unknown Delay, *Journal of Computational and Applied Mathematics*, North-Holland Elsevier, **71**(1996), 67-81.

Burton, T.A. *Volterra Integral and Differential Equations*, Academic Press, Orlando, Florida, 1983.

Caljuk, Z.B. Volterra Integral Equations, *Itogi Nauki, Mat. Analiz*, **15**(1977), 131-198 (in Russian).

Carlson, D.A., An Elementary Proof Of The Maximum Principle Of Optimal Control Problems Governed By A Volterra Integral Equation, *J. Opt. Theory Appl.*, **54**(1987), 43-61.

Casti, J.L. *Alternate Realities: Mathematical Models of Nature and Man*, John Wiley & Sons, New York, 1989.

Cergignani, C., *Theory And Applications Of Boltzmann Equation*, Elsevier, New York, 1975.

Chikrii, A. Guarantee Result in Differential Games with Terminal Payoff, in *Annals of the International Society of Dynamic Games.* New Trends in Dynamic Games and Applications, Birkhauser, **3**(1995), 259-269.

Chikrii, A.A. *Conflict-Controlled Processes*, Kluwer Academic Publ., Dordrecht/ Boston/ London, 1997.

Chikrii A.A. and Eidelman S.D. Game Problems for Fractional Quasilinear Systems, *J. Computers & Mathematics with Applications*, **44**(2002), pp. 835-851.

Clark, C.W. *Bioeconomics Modeling and Fishery Management*, Wiley-Interscience, New York, 1985.

Colton, C., and Kress, R. *Integral Equation Methods In Scattering Theory*, John Wiley, New York, 1983.

Colton, D. and Kress, R. *Inverse Acoustic And Electromagnetic Scattering Theory*, Springer-Verlag, Berlin, 1992.

Connor, M.A. and Hood, D. A Gradient Method for Systems Described by Integral Equation, *IEEE Transac. Automat. Contr.*, **AC-18** (1973), 175-177.

Cooke, K. and Kaplan, J. A Periodicity Threshold Theorem for Epidemics and Population Growth, *Math. Biosci.*, **31**(1976), 87-104.

Copeland, A.C., Ravichandran, G., and Trivedi, M. Localized Radon Transform-Based Detection of Linear Features in Noisy Images, in *Proc. IEEE Conf. on Computer Vision and Pattern Recognition*, Seattle, WA, June 1994, pp. 664-667.

Corduneanu, C. Frequency Domain Criteria For Nuclear Reactor Stability, *Libertas Mathematica*, **1**(1981), 91-116.

Corduneanu, C. *Functional Equations with Causal Operators*, Taylor and Francis, London, 2002.

Corduneanu, C. *Integral Equations And Applications*, Cambridge University Press, New York, 1991.

Costanza, R., Cumberland, J., Daly, H., Goodland R., and Norgaard, R. *An Introduction to Ecological Economics*, Island Press, Washington D.C., 1997.

Courant, R. and Hilbert, D. *Methods of Mathematical Physics*, Volume I, Interscience Publishers, New York, 1953.

Cross, M. and Moscardini A.O. *Learning The Art Of Mathematical Modelling*, Ellis Horwood , Chichester; Halsted Press, New York, 1985.

Cushing, J.D. *Integrodifferential Equations And Delay Models in Population Dynamics*. Lecture Notes in Biomathematics, No.20, Springer-Verlag, Berlin, 1977.

David, P. *Technical Choice, Innovation and Economic Growth*, Cambridge University Press, Cambridge, 1975.

Davies, S. *The Diffusion of Process Innovations*, Cambridge University Press, Cambridge, Cambridge, 1979.

Deans, S. R. *The Radon Transform and Some of Its Applications*, Wiley, New York, 1983.

Delves, L. and Walsh, J. *Numerical Solution of Integral Equations*, University Press, Oxford, 1974.

Dosi, G. *Technical Change and Industrial Transformation*, Macmillan, London, 1984.

Dreyer, T. P. *Modelling With Ordinary Differential Equations*, CRC Press, Boca Raton/Ann Arbor/London/Tokyo, 1993.

Edelstein-Keshet, L. *Mathematical Models in Biology*, Random House, New York, 1988.

Eidelman, S.D. and Chikrii, A.A. Game problems for systems with Volterra evolution. Fractal Games, *Journal of Game Theory and Applications*, Nova Science Publ., VI(2001), 9-44.

Englmann, F.C. Innovation Diffusion, Employment and Wage Policy, *Journal of Evolutionary Economics*, **2**(1992), 179-193.

Eykhoff, P. *System Identification*, Wiley, New York, 1974.

Feistel, R. and Ebeling, W. *Evolution of Complex System*, Kluwer Academic Publishers, Dordrecht, 1989.

Filippi, P., Habault, D., Lefebvre, J.P., and Bergassoli, A., *Acoustics: Basic Physics, Theory and Methods*. Academic Press, San Diego/ London/ Boston/ New York, 1999.

Fisher, J.C. and Pry, R.H. A Simple Substitution Model of Technological Change, *Technological Forecasting and Social Change*, **3**(1971), 75-88.

Fowkes, N.D. and Mahony, J.J. *An Introduction to Mathematical Modelling*. Wiley, New York, 1994

Fowler, A.C. *Mathematical Models in the Applied Sciences*. Cambridge University Press, Cambridge, 1997

Frederick, D.K. and Close, C.M. *Modelling and Analysis of Dynamic Systems*, Houghton Mifflin, Boston, 1978.

Friedman, A. *Industrial Mathematics: A Course In Solving Real-world Problems,* Society for Industrial and Applied Mathematics, Philadelphia,1994.

Friedman, A. *Mathematics in Industrial Problems, Parts I - X*. IMA Volumes in Mathematics and its Applications, Springer-Verlag, New York,1988-1998.

Fulford, G., Forrester, P., and Jones, A. *Modelling With Differential And Difference Equations*, Cambridge University Press, Cambridge / New York, 1997.

Gabasov, R. and Kirillova F. *The Qualitative Theory of Optimal Processes*, Moscow, Nauka Press, 1971 (in Russian).

Galkin, V.A. and Dubovskii, P.B. Solution Of The Coagulation Equation With Unbounded Kernels, *Differential Equations*, **22**(1986), 373-378.

Galperin, E. Deterministic Regression Models for Prediction and Control, *Mathematical Modeling*, **6**(1985), 157-171.

Galperin, E. Global Optimization in Problems with Uncertainties, *Nonlinear Analysis*, **47**(2001), 941-952.

Gindikin, S. (Ed.). *Applied Problems of Radon Transform*, Amer. Math. Soc., Providence, 1994.

Glushkov, V. On a Class of Dynamical Macroeconomic Models, *Upravlyayushchie sistemy i mashiny (Control Systems and Machines)*, Kiev, 1977, No.2, 3-6 (in Russian).

Glushkov, V., Ivanov, V., and Yatsenko, Yu. Analytical Investigation of One Class of Dynamical Models, *Cybernetics*, Plenum Publish. Corp., New York, **16**(1980), No.2, 164-177; **18**(1982), No.3, 390-401.

Golovach, I. and Yatsenko, Yu. Mathematical Modeling of the Process of Replacement and Modernization of the Elements of Manufacturing Systems, *J. Automat. Inform. Sci.*, Scripta Technica Inc., New York, **25**(1992), No.5, 78-81.

Gorjacenko, V. *Stability Methods in the Dynamics of Nuclear Reactor*, Atomizdat, Moscow, 1971 (in Russian).

Goroshko, O.N. and Savin, G.N. *Introduction into The Mechanics of Deformable One-Dimensional Bodies of Variable Length*, Naukova Dumka Press, Kiev, 1971 (in Russian).

Gradshteyn, I. S. and Ryzhik, I. M. *Tables of Integrals, Series, and Products, 6th ed*, Academic Press, San Diego, 2000.

Gripenberg, G., Londen, S.-O., and Staffans, O.J. *Volterra Integral and Functional Equations*, University Press, Cambridge, 1990.

Gurtin, E. and Maccamy, R.G. Non-Linear Age-Dependent Population Dynamics, *Arch. Rat. Mech. Anal.*, **54**(1974), 281-300.

Haberman, R. *Mathematical Models*. Prentice-Hall, New York,1977.

Hale, J.K. *Theory of Functional Differential Equation*, Springer Verlag, New York, 1977.

Hofbauer, J. and Sigmund, K. *The Theory of Evolution and Dynamical Systems*, Cambridge University Press, Cambridge, 1988.

Holmes, E., Lewis, M., Banks, J. and Verr, R. Partial Differential Equations in Ecology: Spatial Interactions and Population Dynamics, *Ecology*, **75**(1994), 17-29.

Hoppensteadt, F.C. and Peskin, S.C. *Mathematics In Medicine And Life Sciences*, Springer-Verlag, New York, 1992.

Hritonenko, N. and Yatsenko, Yu. Integral-Functional Equations for Optimal Renovation Problems, *Optimization*, Gordon and Breach Science Publishers S.A., **36**(1996), 249-261.

Hritonenko, N. and Yatsenko, Yu., *Mathematical Modeling in Economics, Ecology and the Environment*, Kluwer Academic Publishers, Dordrecht /Boston /London, 1999.

Hritonenko, N. and Yatsenko, Yu. *Modeling and Optimization of the Lifetime of Technologies*, Kluwer Academic Publishers, Dordrecht/ Boston/ London, 1996.

Hritonenko, N. and Yatsenko, Yu. Optimization of the Lifetime and Cost of Equipment Under Technological Progress, *J. Automat. Inform. Sci.*, Scripta Technica Inc., New York, **28**(1995), No.2, 117-128.

Hrusa, W.J. A Nonlinear Functional-Differential Equation in Banach Space With Applications To Materials With Fading Memory, *Arch. Rat. Mech. Anal.*, **84**(1984), 99-137.

Hungerbühler, N. Singular Filters for the Radon Backprojection, *J. Appl. Analysis* **5**(1998), 17-33.

Ioffe, A.D., and Tikhomirov, V.D. *Theory of Extremal Problems*, North-Holland, Amsterdam, 1979.

Ivanilov, Yu. and Lotov, A. *Mathematical Models in Economics*, Nauka Press, Moscow, 1984 (in Russian).

Ivanov, V. and Yatsenko, Yu. The Problems of Optimal Control for the Integral Dynamical Models with Controllable Prehistory, *Lecture Notes in Control and Information Sciences*, Springer-Verlag, Berlin, **84**(1986), 350-357.

Ivanov, V., Yanenko, V., and Yatsenko, Yu. Qualitative Study of a Class of Integral Dynamical Models, *Cybernetics*, Plenum Publish. Corp., New York, **19**(1983), 155-167.

Ivanov, V., Yatsenko, Yu., and Galiev, U. Comparison of Some Integral Macroeconomic Models, *Soviet Automatic Control*, New York, **18**(1985), 15-24.

Iwaniec, T. and Sbordone, C. Quasiharmonic Fields, *Ann. I. H. Poincaré – AN*, **18**(2001), 519-572

Jawson, M.A. and Symm G.T., *Integral Equations Methods In Potential Theory And Elastostatics*, Academic Press, New York, 1977.

Jerri, A.J., *Introduction to Integral Equations With Applications*, Marcel Dekker, New York, 1985.

Johansen, L. *Production Functions*, Amsterdam, London, 1972.

Jorgenson, D.W. Capital Theory and Investment Behavior, *Amer. Econ. Rev. (Papers and Proceedings)*, **53**(1963), 247-259.

Kailath, T. *Linear Systems*, Prentice-Hall, Englewood Cliffs, N.J., 1980.

Kantorovich, L. and Gorkov, L. On Some Functional Equations Arising in Analysis of Single-Commodity Economic Model, *Dokl. Akad. Nauk SSSR*, **129**(1959), 732-736 (in Russian).

Kantorovich, L. and Zhiyanov, V. Single-Commodity Dynamic Model of the Economy Allowing for Changes in Asset Structure in the Presence of Technical Progress, *Dokl. Akad. Nauk SSSR*, **211**(1973), 1280-1283 (in Russian).

Kapur, J.N. *Mathematical Modeling*, John Wiley & Sons, New York, 1988.

Karlin, S. *A First Course in Stochastic Processes*, Academic Press, London, New York, 1968.

Kolmogorov A. Qualitative Study of Mathematical Models of Population Dynamics, *Problemy Kibernetiki (Problems of Cybernetics)*, Nauka Press, Moscow, 1972, 3-22 (in Russian).

Kolmogorov, A. and Fomin, S. *The Elements of Function Theory and Functional Analysis*, Nauka Press, Moscow, 1976 (in Russian)

Kunyansky, L. A. Generalized and Attenuated Radon Transforms: Restorative Approach to the Numerical Inversion, *Inverse Problems* **8**(1992), 809-819.

Lakshmikantham, V. Some Problems in Nonlinear Integrodifferential Equations of Volterra Type, *J. Int. Equations*, **10**(1985), 137-146.

Lakshmikantham, V., and Rama Mohana Rao, M. *Theory of Integro-Differential Equations*, Gordon and Breach Science Publishers, 1995.

Lancaster, K. *Mathematical Economics*, Macmillan Comp., New York, 1968.

Landau, L. and Lifschitz, E. *Theory of Elasticity. Course of Theoretical Physics, vol.7*, Pergamon Press, Oxford, 1986.

Lee, E.B. and Markus, L. *Foundations of Optimal Control Theory*, Wiley, New York, 1967.

Legget, R. and Williams, L. Nonzero Solutions of Nonlinear Integral Equations Modeling Infections Diseases, *SIAM J. Math. Anal.*, **33**(1983), No.1, 112-121.

Lenhart, S. and Liao, Y.C., Integro-Differential Equations Associated with Optimal Stopping Time of a Piecewise Deterministic Process, *Stohastics*, **15**(1985), 183-207.

Leontief, W. *Input-Output Economics*, Oxford University Press, New York, 2nd edition, 1986.

Leugering, G., Time Optimal Boundary Controllability Of A Simple Linear Viscoelastic Liquid, *Math. Methods Appl. Sci.*, **9**(1987), 413-430.

Lin, C. C., and Segel L.A. *Mathematics Applied To Deterministic Problems In The Natural Sciences*, Macmillan Publishing Co., New York, 1974.

Lions, J.L. *Sur Quelques Questions D'analyse, de Mecanique et de Controle Optimal*, Presses de l'Universite du Quebec a Montreal, Montreal, 1976.

Ljung, L. *System Identification: Theory for the User*, Prentice-Hall, Englewood Cliffs, N.J., 1987.

Logan, J.D. *Applied Mathematics*, Wiley-Interscience, New York, 1987.

Londen, S.O. and Nohel J.A.. Nonlinear Volterra Integrodifferential Equations Occuring in Heat Flow, *J. Int. Equations*, **6**(1984), 11-50.

Luca, N. A The Stability Of the Solutions of A Class Of Integrodifferential Systems With Infinite Delay, *J. Math. An. Appl*, **67**(1979), 323-339.

Luenberger, D. *Introduction To Dynamic Systems: Theory, Models and Applications*, John Wiley & Sons, New York, 1979.

MacCamy, R. C. A Model For One-dimensional Nonlinear Viscoelasticity, *Quart. Appl. Math.*, **35**(1978), 21-33.

MacCluer, C.R. *Industrial Mathematics: Modeling in Industry, Science and Government*, Prentice-Hall, New York, 2000

MacFarlane, A.G.J. *Dynamical System Models*, Harrap, London, 1970.

Makarov, V., Levin, M., and Rubinov, A. *Mathematical Economic Theory: Pure and Mixed Types of Economic Mechanisms*, North-Holland, 1995 (Advanced Textbooks in Economics, vol. 33).

Malcomson, J.M. Replacement and the Rental Value of Capital Equipment Subject to Obsolescence, *Journal of Economic Theory*, **10**(1975), 24-41.

Mathematical Models for Control of Water Contamination, Mir Press, Moscow, 1981 (in Russian).

May, R.M., Simple Mathematical Models with Very Complicated Dynamics, *Nature*, International Press, **261**(1976), 459-467.

Maz'ya, V.G. *Boundary Integral Equations,* in Analysis IV: Linear And Boundary Integral Equations, Springer-Verlag, Berlin, New York, 1989, 127-233.

McKenzie, L.M. Turnpike Theory, *Econometrica,* **44**(1973), No.5, 841-866.

Medhin, N.G., Optimal Processes Governed By Integral Equations, *J. Math. An. Appl.*, **120**(1986), 1-12.

Mesarovic, M., Macko, D., and Takahara, Y. *Theory of Hierarchical Multilevel Systems*, Academic Press, New York, 1970.

Mikhalevich, V., Volkovich, V., and Yatsenko, Yu. Multicriterial Analysis of Industry Conversion Rates Using Integral Models, *Cybernetics and System Analysis*, Plenum Publish.Corp., New York, **29**(1993), No.1, 26-34.

Miller, R.K. A System Of Volterra Integral Equations Arising In The Theory Of Superfluidity, *An. St. Univ. 'Al. I. Cuza' Iasi,* **19**(1973), 349-364.

Moiseev, N.N., *Models of Ecology and Evolution*, Znanie Press, Moscow, 1983 (in Russian).

Moran, P.A.P. *The Statistical Processes of Evolutionary Theory*, Oxford University Press (Clarendon), London, New York, 1962.

Mordukhovich, B. Coderivatives of Set-Valued Mappings: Calculus and Applications, *Nonlinear Anal. Theory Methods Appl.* **30**(1997), 3059-3070.

Mordukhovich, B. Optimal Control of Difference, Differential and Differential-Difference Inclusions, *J. Math. Sci.*, **100**(2000), 2613-2632

Morse, P.M. and Feshbach H. *Methods of Theoretical Physics*, Volumes I and II, McGraw-Hill Book Company, New York, Toronto, London, 1953.

Murray, J.D. *Mathematical Biology*. Springer-Verlag, Berlin / Heidelberg / New York, 1989.

Nayfeh, A.H. and Mook, D.T. *Nonlinear Oscillations*, Wiley-Interscience, New York, 1979.

Neustadt, L. *Optimization (A Theory Of Necessary Conditions),* Princeton University Press, Princeton, 1976.

Nijkamp, P. *Theory and Application of Environmental Economics*, Elsevier, Amsterdam, 1980.

Nohel, J.A. Nonlinear Conservation Laws With Memory, in: *Computational Method In Science And Engineering, Vol.5,* North Holland, New York, 1987, 269-280.

O'Regan, D. and Meehan, M. *Existence Theory for Nonlinear Integral and Integrodifferential Equations,* Kluwer Academic Publishers, Dordrecht /Boston /London, 1998

Oster, G. The Dynamics of Nonlinear Models With Age Structure, In *Studies in Mathematical Biology. Part II,* Math. Assn. America, Washington. DC, 1978, pp.411-438.

Petrosjan, L.A. and Zakharov, V.V. (Ed.) *Mathematical Models In Environmental Policy Analysis,* Nova Science Publishers, New York, 1997.

Petrov, A. and Pospelov, I. System Analysis of Developing Economics, *Izvestiya Akademii Nauk SSSR. Kibernetika,* 1979, No.2,18-23; No.3, 28-38; No.4, 11-23; No.5, 13-24 (in Russian).

Podowski, M.Z. Nonlinear Stability Analysis For a Class of Differential-Integral Systems Arising From Nonlinear Reactor Dynamics, *IEEE Trans. Aut. Control,* AC-31 (1986), 98-107.

Poincare, H. *New Methods of Celestial Mechanics, Volumes 1-3.* History of modern physics and astronomy, vol. 13, American Institute of Physics, Woodbury, New York, 1993.

Poluektov, R., Pykh, Yu., and Shvytov, I. *Dynamic Models of Ecological Systems,* Gidrometeoizdat, Leningrad, 1980 (in Russian).

Pontryagin, L.S., Boltianskii, V.G., Gamkrelidze, R.V., and Mishchenko, E.F. *The Mathematical Theory of Optimal Processes,* Wiley-Interscience, New York, 1962.

Porter, D. and Stirling, D. *Intergal Equations: a Practical Treatment, From Spectral Theory To Applications,* Cambridge University Press, Cambridge, 1990.

Preinreich, G.A.D. The Economic Life of Industrial Equipment, *Econometrica,* **8**(1940), 12-44.

Primak, A., Kafarov, V. and Kachiashvili, K. *System Analysis of Control of Water Quality,* Naukova Dumka Press, Kiev, 1991 (in Russian).

Prüss, J. *Evolutionary Integral Equations and Applications,* Birkhäuser Verlag, Basel / Boston, 1993.

Pshenichnyi, B.N. *Necessary Conditions For An Extremum,* Marcel Dekker, New York, 1971.

Raafat, F. Survey of Literature on Continuously Deteriorating Inventory Models, *J. Oper. Res. Soc.*, **42**(1991), No.1, 27-37.

Radner, R. Paths of Economic Growth that are Optimal with regard only to Final States: A Turnpike Theorem, *Review of Economic Studies,* **38**(1961), No.76, 1185-1196.

Ramm, A. G. and Katsevich, A. I. *The Radon Transform and Local Tomography*, CRC Press, Boca Raton, 1996.

Ramm, A.G., *Theory And Applications Of Some New Classes Of Integral Equations*, Springer-Verlag, New York, Oxford, 1980.

Renardy, M., Hrusa, W.J., and Nohel, J.A. *Mathematical Problems In Viscoelasticity,* Longman Science Technology, New York, 1987.

Ricker, W.E. Stock and Recruitment, *J. Fish. Res. Bd. Canada,* **11**(1954), 559-623.

Rockafeller, T. *Convex Analysis*, Princeton University Press, Princeton, 1970.

Rorres, C. Stability of an Age Specific Population with Density Dependent Fertility, *Theor. Pop. Biol.*, **10**(1976), 26-46.

Rubinov, A. M. *Abstract Convexity and Global Optimization*, Kluwer Academic Publishers, Dordrecht /Boston /London, 2000.

Saaty, T.L. and Joyce, M. *Thinking With Models: Mathematical Models In the Physical, Biological, And Social Sciences,* Pergamon, Oxford, New York, 1981.

Shaikhet, L.E., On The Optimal Control Of Integral-Functional Equations, *Prikl. Mat. Mekh. USSR.*, **49**(1985), 704-712.

Sharkovsky, A.N., Maistrenko, Yu.L. and Romanenko, E.Yu., *Difference Equations and Their Applications*, Kluwer Academic Publishers, Dordrecht /Boston /London, 1994.

Sharpe, F.R. and Lotka, A.J. A Problem in Age-Distribution, *Philosophical Mag.*, **21**(1911), 435-438.

Sheu, S-H. A Generalized Block Replacement Policy with Minimal Repair and General Random Repair Costs for a Multi-Unit System, *J. Oper. Res. Soc.*, **42**(1991), No.4, 331-341.

Silverberg, G. and Lehnert, D. Long Waves and Evolutionary Chaos in a Simple Schumpeterian Model of Embodied Technical Progress, *Structural Change and Economic Dynamics*, **4**(1993), No.1, 9-37.

Silverberg, G. Technical Progress, Capital Accumulation and Effective Demand: A Self-Organizing Model, in *Economic Evolution and Structural Adjustment*, Springer Verlag, Berlin, Heidelberg, New York, Tokyo, 1987.

Silverberg, G., Dosi, G., and Orsenigo, L. Innovation, Diversity and Diffusion: A Self-Organizing Model, *The Economic Journal*, **98**(1988), 1032-1054.

Smith, V. *Investment and Production: A Study in the Theory of the Capital-Using Enterprise*, Harvard University Press, Cambridge, Massachusetts, 1961.

Smoller, J. *Shock Waves and Reaction-Diffusion Equations*, Springer-Verlag, New York, 1983.

Soderstrom, T. and Stoica, P. *System Identification*, Prentice-Hal Englewood Cliffs, N.J., 1987.

Solow, R. Investment and Technical Progress, in *Mathematical Methods in the Social Science*, Stanford Univ. Press, 1960, 89-104.

Solow, R. Technical Change and the Aggregate Production Function, *Rev. of Economics and Statistics*, **XXXIX**(1957), 312-330.

Staffans, O.J. On the Almost Periodicity of the Solutions of An Integrodifferential Equation, *J. Int. Equations*, **8**(1985), 249-260.

Stakgold, I. *Green's Functions And Boundary Value Problem*, Wiley-Interscience, New York, 1979. d in

Stanislav, J. F. *Mathematical Modeling Of Transport Phenomena Processes,* Ann Arbor Science, Ann Arbor, , 1982.

Stoneman, P. *The Economic Analysis of Technological Change*, Oxford University Press, Oxford, 1983.

Strang, G. and Temam, R. A Problem In Capillarity and Plasticity, *Mathematical Programming Study*, North-Holland Publ. Co., **17**(1982), 91-102.

Streifer, W. Realistic Models in Population Ecology, *Advances in Ecological Research*, **8**(1974), 200-266.

Svobodny, T. *Mathematical Modeling for Industry and Engineering*, Prentice-Hall, New Jersey, 1998.

Swick, K.E. A Nonlinear Age-dependent Model of Single Species Population Dynamics, *SIAM Journal Appl. Math*, **32**(1977), No.2, 484-498.

Sydow, A., Lux, T., Rosé, H., Rufeger, W., and Walter, B. Conceptual Design of the Branch-Oriented Simulation System DYMOS (Dynamic Models for Smog Analysis), *Transactions of the Society for Computer Simulation International*, **15**(1998), No. 3, 95-100

System Analysis and Methods of Mathematical Modelling in Ecology, Cybernetics Institute of Ukrainian Acad. Sci., Kiev, 1990 (in Russian).

Tabrizi, M.H.N., Copperthwaite, L., and Billings, S.A. Influent Flow Modelling Using Identification Techniques, *International Journal of Modelling And Simulation*, ACTA Press, Calgary, **21**(2001), No.1, 3-22

Tijms, H.C. *Stochastic Modelling and Analysis*, Wiley, New York, 1986.

Tinbergen, J. and Bos, H.C. *Mathematical Models of Economic Growth*, McGraw-Hill Book Company, New York, London, 1962.

Tsokos, C., and Padgett, W. *Random Integral Equations with Applications to Life Sciences And Engineering*, Academic Press, New York, 1974.

Vasiliev, F.P. *Solution Methods to Extremal Problems*, Nauka, Moscow, 1981 (in Russian).

Vierck, R. *Vibration Analysis*, International Textbook Company, Scranton, 1967.

Vinokurov, V.R., Optimal Control Of Processes Described By Integral Equations, I, II, III, *SIAM J. Control*, 7(1969), 324-355.

Volterra, V. *Lecons sur la Theorie Mathematique de la Lutte Pour la Vie*, Gauthier-Villars, Paris, 1931.

Volterra, V. *Theory of Functionals and of Integral and Integro-Differential Equations*, Dover Publications, New York, 1959.

Väth, M. *Volterra and Integral Equations of Vector Functions*, Marcel Dekker, New York, 2000.

Warga, J. *Optimal Control Of Differential And Functional Equations*, Academic Press, New York, 1972.

Webb, G.F. *Theory of Nonlinear Age-Dependent Population Dynamics*, M. Dekker, New York, Basel, 1985.

Whitham, G.B. *Linear and Nonlinear Waves*, Wiley-Interscience, New York, 1974.

Yatsenko, Yu. and Hritonenko, N. *Integral Models of Developing Systems: Some Theoretical and Numerical Aspects*, Preprint 94-1 of the Cybernetics Institute of Ukrainian Acad. Sci., Kiev, 1994.

Yatsenko, Yu. and Hritonenko, N. Optimization in Integral Model of Developing Systems, *Optimization*, Gordon and Breach Science Publishers S.A., **31**(1994), 179-192.

Yatsenko, Yu. Application of Integral Dynamical Models in Mathematical Economics, *Visnyk AN USSR*, Kiev, 1986, No.9, 25-30 (in Ukrainian).

Yatsenko, Yu. *Integral Models of Systems with Controllable Memory*, Naukova Dumka Press, Kiev, 1991 (in Russian).

Yatsenko, Yu. Integral Volterra Equations with Variable-Duration After-Effect, *Soviet J. Automat. Inform. Sci.*, Scripta Technica Inc., New York, **24**(1991), No.4, 25-31.

Yatsenko, Yu. On Systems of Nonlinear Volterra Integral Equations with Unknown Lower Limit of Integration, *Ukrainian Mathematical Journal*, New York, Plenum Publish.Corp., **38**(1986), No.1, 35-42.

Yatsenko, Yu. Simulation of Certain Oscillatory Biological Processes, *Cybernetics*, New York, Plenum Publish.Corp., **14**(1978), No.5, 529-535.

Yatsenko, Yu. Volterra Integral Equations with Unknown Delay Time, *Methods and Application of Analysis,* International Press, **2**(1995), 408-419.

Young, L.C. *Lectures on the Calculus of Variation and Optimal Control Theory*, Saunders, Philadelphia, Pa., 1969.

Yu., D., *Natural Boundary Integral Method and Its Applications*, Kluwer Academic Publishers, Dordrecht /Boston /London, 2002.

Zabreiko, P.P., Koshelev, A.Z., Krasnoselskii, M.A., Mikhlin, S.G., Rakovshchik, L.S., and Statsenko, V.Ya. *Integral Equations – A Reference Text,* Noordhoff International Publishers, Leyden, 1975.

Index